INTRODUCTION:

Why renewable energy, science and technical skills?

High schools struggle to get and keep students engaged in the study of science, while industry struggles to attract employees with advanced technical skills. As construction trades and science teachers, we see a great opportunity to combine the growing national interest in renewable energy with lab science and hands-on skills to provide a truly integrated, contextual curriculum to engage students.

- Renewable energy provides a political, economic and technical framework for the study of scientific concepts and methodology.
- Renewable energy utilization rests on the development of advanced technical skills: engineering research and design; electrical power production, transmission and utilization; manufacturing; transportation modeling; urban planning and design among others.
- The translating of scientific concepts into working physical models offers unparalleled opportunities for students to practice creative and critical thinking, and to problem-solve in a tangible context.

How does it work?

- Designed to be used by either academic or career and tech ed teachers.
- Can be taught as a full year academic science class (see the referenced Colorado Model Content Standards and National Science Education Standards for each unit).
- Individual units, design briefs, or activities can be used as supplemental modules for traditional science or construction trades classes.
- Works as the basis of an after school enrichment club.
- Can be taught in a shop or science lab setting.

What is included?

- 10 Instructional Units plus a Final Project: More than enough for a full academic year: choose the units, design briefs and activities which work best for your students.
- Instructional content on renewable energy and science topics
  - Power points, videos, website resources, vocabulary, handouts
- Design Briefs: student designed and constructed projects
  - Instructions, material requirements, PDF format assignments and grading matrices
- National Science Education Standards Matrix
- Supplemental resources and activities list

About the Design Briefs:

A Design Brief poses a life situation problem which students need to solve using a design, build, test and repeat format. Typically, a combination of individual and team work is required to develop the solution. Students are given topical background, a goal, and then allowed time for research and development of a solution. Requirements for individual research, group negotiations to agree on a common solution, teamwork to produce a working model of the solution, and group presentations all provide students with exposure to "real-life" work place skills in addition to academic content – a truly integrated, contextual approach to science and technical instruction.

Design Briefs are presented in two formats:
- In one, students determine the steps required to meet all criteria and accomplish the goal listed in the Brief. This is considered a project based design brief.
- In the second, the Brief outlines specific steps for an experiment that students will perform, resulting in a more structured experience.

Please consider the Design Briefs as a starting point for your class experience, as you may need to modify some of the activities to fit your individual needs. For example: We do the testing of the Fizzi Rockets outside against our school building, approximating the distance based on the number of levels of brick. You may not have a place on your building to do this and have to find a different method of measuring height.

Each Design Brief includes checklists and grading rubrics for use by both teacher and students, so please take a few minutes to review the materials on the CD.

You will find that your students enjoy the challenge of working in teams to solve a problem in a very hands-on format which reinforces their learning in your classroom.

Table of Contents

Fall semester designed for 82 45-minute class periods.
Spring semester designed for 84 45-minute class periods.

FALL SEMESTER

SPRING SEMESTER

UNIT 6: HYDROGEN AND FUEL CELLS
Twenty, 45 minute class periods or 10 block class periods.

6.1. Introduction to hydrogen
6.2. Build Your Own Fuel Cell Project
6.3. The Fuel Cell Invention Design Brief

UNIT 7: SUSTAINABLE TRANSPORTATION
Twenty, 45 minute class periods or 10 block class periods.

7.1. Introduction to sustainable transportation
7.2. The Fuel Cell Vehicle Design Brief

UNIT 8: BIOMASS ENERGY AND BIO-FUELS
Eighteen, 45 minute class periods or 9 block class periods.

8.1. Introduction to biomass energy
8.2. Biodiesel

Supplemental Activities:
8.3. No Fossils in This Fuel, Making Ethanol Project
8.4. Build Your Own Biogas Generator

UNIT 9: GEOTHERMAL ENERGY
Ten, 45 minute class periods or 5 block class periods.

9.1. Introduction to geothermal energy
9.2. Steam turbine project
9.3. Geothermal building systems
9.4. Geothermal workbook exercises
9.5. Model Geyser Project

UNIT 10: HYDROPOWER
Six, 45 minute class periods or 3 block class periods.

10.1. Introduction to hydroelectric energy
10.2. Calculating the energy in water—field trip
10.3. Generate Your Own Hydropower activity
10.4. Ocean power

UNIT 11: RENEWABLE ENERGY FINAL PROJECT
Ten, 45 minute class periods or 5 block class periods.

11.1. Final option 1
11.2   Final option 2

APPENDIX

Fall Semester

UNIT 1:

Energy Science, Fossil Fuels, & Climate Change

UNIT OBJECTIVES:
- Increase student awareness of current energy technologies and climate change.
- Introduce students to different forms of energy.
- Prepare students to work safely with power tools and electricity.
- Increase student understanding of their contribution to global climate change.

VOCABULARY:

Peak oil	Entropy	Ohms
Climate change	Greenhouse effect	Kinetic energy
Fossil fuels	Amps	Potential energy
Laws of thermodynamics	Volts	

COLORADO MODEL CONTENT STANDARDS: BENCHMARKS FOR GRADES 9-12

1.2 Select and use appropriate technologies to gather, process, and analyze data to report information related to an investigation.

1.4 Recognizing and analyzing alternative explanations and models.

1.6 Communicate and evaluate scientific thinking that leads to particular conclusions

2.4. Word and chemical equations are used to relate observed changes in matter to its composition and structure (for example: conservation of matter).

2.5. Quantitative relationships involved with thermal energy can be identified, measured, calculated and analyzed (for example: heat transfer in a system involving mass, specific heat, and change in temperature of matter).

2.6. Energy can be transferred through a variety of mechanisms and in any change some energy is lost as heat (for example: conduction, convection, radiation, motion, electricity, chemical bonding changes).

3.6. Changes in an ecosystem can affect biodiversity and biodiversity contributes to an ecosystem's dynamic equilibrium.

3.7. There is a cycling of matter (for example: carbon, nitrogen) and the movement and change of energy through the ecosystem (for example: some energy dissipates as heat as it is transferred through a food web).

4.4. There are costs, benefits, and consequences of natural resource exploration, development, and consumption (*for example: geosphere, biosphere, hydrosphere, atmosphere and greenhouse gas*).

4.5.There are consequences for the use of renewable and nonrenewable resources.

5.1 Print and visual media can be evaluated for scientific evidence, bias and opinion

5.2 Identify reasons why consensus and peer review are essential to the Scientific Process.

5.3 Graphs, equations, or other models are used to analyze systems involving change and constancy (for example, comparing the geologic time scale to shorter time frames, exponential growth, a mathematical expression for gas behavior; constructing a closed ecosystem such as an aquarium).

5.4 There are cause - effect relationships within systems (for example: the effect of temperature on gas volume, effect of CO_2 level on the greenhouse effect, effects of changing in nutrients at the base of a food pyramid).

5.5 Scientific knowledge changes and accumulates over time; usually the changes that take place are small modifications of prior knowledge about major shifts in the scientific view of how the world works do occur.

5.6 Interrelationships among science, technology and human activity lead to further discoveries that impact the world in positive and negative ways.

5.7 There is a difference between scientific theory and scientific hypothesis.

UNIT 1 TEACHER SCHEDULE

UNIT TIME PERIOD:
This unit is designed to take 16 45-minute class periods.

1.1. FIRST-DAY WELCOME
Day 1
Materials: District Safety Contract

1.2. ELECTRICAL AND CHEMICAL SAFETY
Day 2
Materials: Electrical Safety PowerPoint Presentation, District Safety Test

1.3. INTRODUCTION TO ENERGY
Day 3

1.4. *FIZZI ROCKET DESIGN BRIEF*
Day 4: Introduction to the *Fizzi Rocket Design Brief*
Materials: Copy of Fizzi Rocket Design Brief for each students, empty film canisters.
Days 5-10: Design, construction, and preliminary testing
Materials:
- *Alka-Seltzer*
- *Cups for water*
- *Safety glasses for all students*
- *Assorted methods for fastening—wood glue, hot glue, screws, nails*
- *Basic shop tools—hammers, screw drivers, utility knives (safety version, self-retracting), tape measures,*
- *2"-18" thermometers (one for each team)*
- *Assorted materials for rocket construction—balsa wood, cardboard, paper, etc.*
- *Students may bring their own materials to use if approved by the teacher.*
Days 11-12: Final testing and presentations

1.5. THE OIL STORY, NUCLEAR ENERGY, AND FOSSIL FUEL POSTER PROJECT
Days: 13-14: Oil Story, nuclear energy, and poster research and creation
Materials: Peak Oil Story PowerPoint, copies of the NEED Fossil Fuel handouts and Energy for Keeps Nuclear handout. Access to materials to create posters (computers, large format printer or regular printer and large paper, glue, scissors, etc).
Day 15: Fossil fuel poster presentations

1.6. CLIMATE CHANGE AND INTRODUCTION TO RENEWABLE ENERGY
Day 16
Materials: Tackling Climate Change in the U.S. PowerPoint, Your Future and Its Stewardship PowerPoint

1.1. FIRST DAY WELCOME

Resources & Materials Needed:
- *District Safety Contract*

INTRODUCTION TO THE COURSE
Provide students with an introduction to the course, describing the topics to be covered, schedule for the year, grading system, etc.

HAND TOOL AND SHOP SAFETY
Give students a tour of the shop and show them the tools and equipment that they might use over the course of the year. Discuss general shop and tool safety and any safety rules and procedures that they will need to follow during the course.

Provide each student with a copy of the district safety contract and guidelines.

1.2. ELECTRICAL & CHEMICAL SAFETY

Resources & Materials Needed:
- *Electrical Safety* PowerPoint presentation.
- Safety Test

Studying and working with energy often involves dealing with electricity and various chemicals. Students need to be aware of the importance of safe behavior at all times. Make sure students now what to do in the case of accidents, injuries, fires, or other emergency situations.

ELECTRICAL SAFETY PRESENTATION
Go through the electrical safety PowerPoint presentation with student and cover all topics thoroughly. The presentation covers:
- Electrical hazards
- Basic safety requirements and procedures
- Safety related equipment

CHEMICAL SAFETY
Review the safety issues and procedures of working with chemicals, particularly NAOH and Methanol.

Give students Safety Test to insure that they understand safety information.

1.3. INTRODUCTION TO ENERGY

The knowledge of the basics types of energy and the law of thermodynamics are a necessary foundation for fully understanding the topics presented throughout the course. Present the information in section 1.3 below to students and make sure they are comfortable with the information before proceeding.

THE LAWS OF THERMODYNAMICS
Energy is the ability to bring about change or to do work, and thermodynamics is the study of energy. There are two basic laws of thermodynamics that rule how energy behaves and can be used:

First Law of Thermodynamics: Energy can be changed from one form to another, but it cannot be created or destroyed. The total amount of energy and matter in the Universe remains constant, merely changing from one form to another. The First Law of Thermodynamics (Conservation) states that energy is always conserved, it cannot be created or destroyed. In essence, energy can be converted from one form into another.

The Second Law of Thermodynamics states that "in all energy exchanges, if no energy enters or leaves the system, the potential energy of the state will always be less than that of the initial state." This is also commonly referred to as **entropy.** A spring-driven watch will run until the potential energy in the spring is converted, and not again until energy is reapplied to the spring to rewind it. A car that has run out of gas will not run again until you walk 10 miles to a gas station and refuel the car. Once the potential energy locked in carbohydrates is converted into kinetic energy (energy in use or motion), the organism will get no more until energy is input again. In the process of energy transfer, some energy will dissipate as heat. **Entropy** is a measure of disorder: cells are NOT disordered and so have low entropy. The flow of energy maintains order and life. Entropy wins when organisms cease to take in energy and die.

KINETIC AND POTENTIAL ENERGY FORMS
Energy exists in many forms, such as heat, light, chemical energy, and electrical energy. These forms can be grouped into two types, kinetic and potential:

Kinetic Energy is motion—of waves, electrons, atoms, molecules, substances, and objects.

Electrical Energy is the movement of electrical charges. Everything is made of tiny particles called atoms. Atoms are made of even smaller particles called electrons, protons, and neutrons. Applying a force can make some of the electrons move. Electrical charges moving through a wire is called electricity. Lightning is another example of electrical energy.

Radiant Energy is electromagnetic energy that travels in transverse waves. Radiant energy includes visible light, x-rays, gamma rays and radio waves. Light is one type of radiant energy. Solar energy is an example of radiant energy.

Thermal Energy, or heat, is the internal energy in substances—the vibration and movement of the atoms and molecules within substances. Geothermal energy is an example of thermal energy.

Motion Energy is the movement of objects and substances from one place to another. Objects and substances move when a force is applied according to Newton's Laws of Motion. Wind is an example of motion energy.

Sound is the movement of energy through substances in longitudinal (compression/rarefaction) waves. Sound is produced when a force causes an object or substance to vibrate—the energy is transferred through the substance in a wave.

Potential Energy is stored energy and the energy of position—gravitational energy. There are several forms of potential energy.

Chemical Energy is energy stored in the bonds of atoms and molecules. It is the energy that holds these particles together. Biomass, petroleum, natural gas, and propane are examples of stored chemical energy.

Stored Mechanical Energy is energy stored in objects by the application of a force. Compressed springs and stretched rubber bands are examples of stored mechanical energy.

Nuclear Energy is energy stored in the nucleus of an atom—the energy that holds the nucleus together. The energy can be released when the nuclei are combined or split apart. Nuclear power plants split the nuclei of uranium atoms in a process called fission. The sun combines the nuclei of hydrogen atoms in a process called fusion. Scientists are working on creating fusion energy on earth, so that someday there might be fusion power plants.

Gravitational Energy is the energy of position or place. A rock resting at the top of a hill contains gravitational potential energy. Hydropower, such as water in a reservoir behind a dam, is an example of gravitational potential energy.

1.4. *FIZZI ROCKET DESIGN BRIEF*

INTRODUCTION

Resources & Materials Needed:
* Copies of the *Fizzi Rocket Design Brief* for each students, and empty film canisters.

The *Fizzi Rocket Design Brief* has students design, build and test chemical-powered rockets, with the goal of designing a rocket to reach the highest altitude possible. **See the design brief for full details and instructions.**

HAND OUT THE DESIGN BRIEF:
* Break students into groups of two to four.
* Have students put their name on the front cover.
* Hand out film canisters.
* Assign due dates and have the students fill them into the appropriate spaces.
* Carefully go through the requirements, specifications and restrictions. Students do not write on the grading rubric on the last page. Make sure that everyone understands that deviating from or misinterpreting these requirements will affect their final grade.

REVIEW THE DESIGN BRIEF PROCEDURES:
* Go over the activity procedures and the sections of the design brief.
* Review the rules for the design brief and the schedule and guidelines for design, testing, and final evaluation.
* Students should begin their design brief and recording ideas for rocket designs.

DESIGN, CONSTRUCTION, & PRELIMINARY TESTING

Resources & Materials Needed:
* Alka-Seltzer

* Safety glasses for all students
* Assorted methods for fastening—wood glue, hot glue, screws, nails
* Basic shop tools—hammers, screw drivers, utility knives (safety version, self-retracting), tape measures
* 2"-18" thermometers (one for each team)
* Assorted materials for rocket construction—balsa wood, cardboard, paper, etc.
* Students may bring their own materials to use if approved by the teacher.

DESIGN

Students need to have finalized their initial rocket designs by this time. Review each team's final design.

> **Check point 1 - No group may progress beyond Final Design without this step being signed off by the instructor.**

CONSTRUCTION AND PRELIMINARY TESTING:

During this period students construct, test, and modify their designs:

- Students begin the construction phase
- Students work on completing the sections of their design brief as appropriate.
- Students begin testing their Rocket and making changes as necessary to achieve the highest altitude. It is not unusual for rockets to achieve altitudes of 50'-100' (you can use a building or flag pole as a reference to measure altitudes reached).
- Teams prepare for final testing.

Depending on the number of teams, you may want to shorten this period and reserve an additional day for final rocket testing. You may also want to give the day-fourteen climate change presentation and assignment mid way through the testing period.

FINAL TESTING & PRESENTATIONS

> **Resources & Materials Needed:**
> - Alka-Seltzer
> - Cups for water
> - Safety glasses for all students

FINAL TESTING:

Supervise and evaluate the final testing of each team's rocket:

- Students need to have their design briefs, rockets and fuel (Alka-Seltzer & water) ready for testing and evaluation.
- Make sure that students accurately record the results in Part 5 of their design brief.
- Instruct students to finish the final sections of their design briefs and prepare to present their results to the rest of the class.

PRESENTATIONS:

Have each team give a short presentation on their rocket, preliminary testing, and final results. Students turn in their design briefs.

1.5. THE OIL STORY, NUCLEAR ENERGY, AND FOSSIL FUEL POSTER PROJECT

Resources & Materials Needed:
- Copies of the NEED Fossil Fuel handouts.
- *Energy for Keeps* Nuclear handouts.
- Access to materials to create posters (computers, large format printer or regular printer and large paper, glue, scissors, etc).
- *The Peak Oil Story* PowerPoint presentation.

THE OIL STORY:

Present *The Peak Oil Story and the Importance of Community Development* PowerPoint, by Roger Taylor, NREL . The presentation introduces students to the concept of peak oil and some of the issues of oil supply and demand.

NUCLEAR ENERGY:

Nuclear energy is a controversial energy source. It is not a renewable energy source, but because it is a technology not based an fossil fuels many people think nuclear power plants could play an important role in reducing carbon emissions and battling climate change. However, many others feel the risk of accidents and the issues of storing nuclear waste for thousands of years are too significant to warrant the development of this energy source.

Review with students the handout information and the pros and cons of nuclear energy. Ask them what they think should be the role of nuclear energy—using their knowledge of both the potential and barriers of renewable energy technologies, as well as the consequences and limitations of fossil fuel use.

POSTER RESEARCH AND CREATION:

Students will spend two days developing posters on a commonly used fossil fuel.

- Break class into four groups and hand each group one of the four fossil fuels handouts. Coal, Petroleum, Natural Gas, and Propane.
- Each group is to create a poster of their topic, including but not limited to the following:
  - Title and basic introduction of topic
  - Uses
  - Drawbacks
  - Supply

On day six each group should present their poster to the rest of the class by hanging them around the room.

POSTER PRESENTATIONS:
Each group should present their poster to the rest of the class:
- Have students hang their posters around the room.
- Groups should then rotate from poster to poster with sticky notes for each student in the group. Each group should spend no more than 5 minutes at each poster. Students should write questions or comments that they have on the sticky notes and stick them to the poster they are critiquing.
- Students may then go back to their poster and read the critiques and questions and prepare a short presentation to the class, outlining their topic and addressing the questions and critiques.

1.6. CLIMATE CHANGE AND INTRODUCTION TO RENEWABLE ENERGY

Resources & Materials Needed:
- *Tackling Climate Change in the U.S.* PowerPoint presentation.
- *Your Future and Its Stewardship* PowerPoint presentation.

Present *Tackling Climate Change in the U.S.: The Potential for Energy Efficiency and Renewable Energy* (by Chuck Kutscher, NREL) PowerPoint and *Your Future and Its Stewardship* PowerPoint, by Brent Nelson.

CLIMATE CHANGE CURRENT EVENTS PAPER ASSIGNMENT:
Assign students a current events research paper project, with the following guidelines:
- Research an current news article on global warming.
- Write a 1-2 page paper discussing whether the article was for or against global warming, whether you agree with it or not and why.
- Provide supporting material or references for your conclusions when appropriate.

UNIT 2:

Exploring Home Energy Use and Conservation

UNIT OBJECTIVES:
- Increase awareness of home energy use.
- Understand and discuss what forms of energy are used in homes.
- Apply knowledge of energy use, conservation and efficiency to real life situations.
- Increase awareness of a carbon footprint and what can be done to reduce it.

VOCABULARY:

Energy	Energy Star	BTU
Energy conservation	KWh	Carbon footprint
Fuel	Climate	Heating/cooling degree day

COLORADO MODEL CONTENT STANDARDS: BENCHMARKS FOR GRADES 9-12

1.1 Ask questions and state hypotheses, using prior scientific knowledge to help design and guide development and implementation of a scientific investigation.

1.2 Select and use appropriate technologies to gather, process, and analyze data to report information related to an investigation.

1.3 Identify major sources of error or uncertainty within an investigation. (for example: particular measuring devices and experimental procedures.

1.4 Recognizing and analyzing alternative explanations and models.

1.6 Communicate and evaluate scientific thinking that leads to particular conclusions.

2.5. Quantitative relationships involved with thermal energy can be identified, measured, calculated and analyzed (for example: heat transfer in a system involving mass, specific heat, and change in temperature of matter).

2.6. Energy can be transferred through a variety of mechanisms and in any change some energy is lost as heat (for example: conduction, convection, radiation, motion, electricity, chemical bonding changes).

2.8. Quantities that demonstrate conservation of mass and conservation of energy in physical interactions can be measured and calculated.

4.4. There are costs, benefits, and consequences of natural resource exploration, development, and consumption (*for example: geosphere, biosphere, hydrosphere, atmosphere and greenhouse gas*).

4.5. There are consequences for the use of renewable and nonrenewable resources.

5.3 Graphs, equations, or other models are used to analyze systems involving change and constancy (for example, comparing the geologic time scale to shorter time frames, exponential growth, a mathematical expression for gas behavior; constructing a closed ecosystem such as an aquarium).

5.4 There are cause - effect relationships within systems (for example: the effect of temperature on gas volume, effect of CO_2 level on the greenhouse effect, effects of changing in nutrients at the base of a food pyramid).

5.6 Interrelationships among science, technology and human activity lead to further discoveries that impact the world in positive and negative ways.

UNIT 2 TEACHER SCHEDULE

UNIT TIME PERIOD:
This unit is designed to take 10 45-minute class periods, coinciding with a 7-day period for at-home student research.

2.1. INTRODUCTION TO ENERGY CONSERVATION
Day 1: Introduction to energy conservation and the *Personal Energy Audit Design Brief*
Materials: *A copy of the Personal Energy Audit for each student, computers for web research of conversions from kWh to tons of CO_2.*

2.2. APPLIANCE ACTIVITY
Day 2: "Scavenger Hunt" and/or "Watt's Up?"
Materials:
- *Access to appliances and tools around the school and/or an assortment of household electrical devices brought into class.*
- *Kill-a-Watt or other watt meter (optional)*
- *A copy for each student of the semester reading ("Affluenza" by John de Graaf, David Wann, and Thomas Naylor).*

2.3. CHOICE OF ACTIVITIES
DAYS 3-8: Some or all of the additional activities:
School Building Survey
Energy Action Project
Lighting Activity
Materials: *See individual activity guides for the materials needed for each activity.*

2.4.: *Personal Energy Audit* completion and presentations
Days 9-10

2.1. INTRODUCTION TO ENERGY CONSERVATION AND THE *PERSONAL ENERGY AUDIT DESIGN BRIEF*

Resources & Materials Needed:
- A copy of the *Personal Energy Audit* for each student.

INTRODUCTION TO ENERGY CONSERVATION

Note to the Teacher: The *Personal Energy Audit Design Brief* will ask your students to evaluate a number of energy concepts. Use the information provided below to create a presentation that explains many energy concepts before you distribute the *Personal Energy Audits Design Brief* to your students.

CONSERVATION AND EFFICIENCY:

Discuss with students the difference between energy conservation and energy efficiency:

Efficiency is using less energy to do the same amount of useful work. Compact florescent lamps (CFLs) are a good example of efficiency, since they use significantly less energy to produce the same amount of light as an old-fashioned incandescent light bulb.

Conservation is finding ways to save energy by reducing or eliminating the activities that are responsible for energy use. Turning off some or all of the lights in a room when less light is needed or no one is around would be examples of conservation.

Have students brainstorm different examples of conservation and efficiency.

Some additional examples could be:
- More efficient cars like hybrids vs. avoiding unnecessary car trips, carpooling, or riding a bike
- A high-efficiency water heater vs. using less hot water by taking shorter showers or washing laundry on colder settings.
- An Energy Star furnace vs. turning down the temperature on the thermostat.

Students will be able to explore efficiency in more detail in unit 3.

ENERGY USE IN THE HOME:
There are many sources and types of energy, and many homes in the U.S. use at least two forms: electricity, plus a fuel for heating needs.

Electricity is an extremely versatile form of energy, and is capable of supplying the energy for almost any household demand: lights, computers, TVs, refrigerators, stoves, air conditioners, water heaters, furnaces--a few people even have electric cars.

Heating fuels are energy forms that people burn in devices in their homes—typically for hot water and space heating, and often cooking as well. In Colorado, the main heating fuel people use in their homes is natural gas (methane), though some people use wood or propane as well, especially in rural areas. In other areas, fuel oil is also commonly used.

You may also want to discuss with students other forms of energy that people use in their lives, such as transportation fuels and direct solar energy.

Ask students about what forms of energy are used in their homes and for which applications.

DIRECT AND INDIRECT ENERGY USE:
An important concept to consider when dealing with conservation and efficiency is the difference between using a primary energy source directly or indirectly. When we use natural gas or wood to create heat in our homes, for example, we are directly burning a primary energy source.

The electricity we use in our homes is a secondary energy form that has to be generated by converting other primary forms of energy—such as sunlight, coal, wind, or natural gas. In the process of conversion not all the energy contained in the fuel can be converted entirely to electricity, and the remainder is typically lost as heat. For conventional coal or natural gas power plants, these heat losses are often around 2/3 of the total energy content of the original fuel—which means it usually takes about 3 kilowatt-hours (kWh) of coal or natural gas energy to generate 1kWh of electricity.

This is why purchasing electricity is much more expensive than natural gas. In Colorado:

For residential Xcel customers, if the total cost of kWh (including all services charges and taxes) is $0.14, that is about $4 per 100,000 Btu.

- A therm of natural gas costs about $0.98 per therm, which is less that $1 per 100,000 Btu.

So, in Colorado, energy in the form of electricity costs about 4 times as much as energy in the form of natural gas. This illustrates why many homes use natural gas for space heating and hot water.

ENERGY UNITS:

To be able to fully understand energy science and technology, and to be able to quantify energy production and use, students need to understand the different units used to measure energy, and to be able to convert figures from one unit system to another.

The British Thermal Unit (Btu) is defined as the amount of heat necessary to raise the temperature of one pound of water one degree Fahrenheit. The Btu is a standard unit of heat that is used extensively throughout the U.S.

The following chart, found on page 3 of the *Audit*, provides students with Btu value of other common units of energy:

Amount	Btu Equivalent
1 kwh electricity	3,413 Btu
1 cu. Ft. natural gas	750 Btu
1 gallon #2 fuel oil	89,700 Btu
1 gallon bottled gas	73,200 Btu
1 pound bottled gas	17,264 Btu
1 full cord of hardwood	11,500,000 Btu
1 full cord softwood	6,500,000 Btu

Energy utilities commonly measure natural gas in therms, with one therm equal to 100,000 Btu.

Electrical energy is typically measured in kilowatt-hour (kWh), while power consumption is most commonly measured in Watts (W).

It is important for students to understand and be comfortable dealing with power and energy, particularly watts and kilowatt-hours. Energy use is equal to power multiplied by time, which can be written:

$$(power) \times (time) = Energy$$

For example, the wattage rating on a light bulb or appliance is a measure of power, and only after multiplying by the duration of time in use can the total powered consumed be determined. If a 100W light bulb is turned on for 10 hours, then we can calculate:

$$100W \times 10hrs = 1000Wh = 1kWh$$

BUT HOW MUCH IS A kWh, REALLY?

One way to better understand the energy use in our homes is to compare it to the energy we use to fuel our own bodies.

The energy contained in food is typically measured in calories. A calorie is the amount of energy needed to increase the temperature of one gram of water one degree C. However, what most people think of as a "calorie" --the units used on food packaging-- are in fact kilocalories (kcal) or "large calories," equal to 1000 calories. 1 kcal is equal to 3.968 Btu. Knowing this, students can covert their daily food intake into other units of energy. Assuming a 2,000 "calorie" (really kilocalorie) daily food intake, the amount of energy in kWh that a person uses in 24 hours can be calculated:

$$(2,000 kcal) \times (3.986 Btu/kcal) \times (1 kWh/3412 Btu) = 2.3 kWh$$

This can be compared to the energy used by a 100W incandescent light bulb over the same 24-hour period:

$$(100W) \times (24hrs) = 2.4 kWh$$

So, a human actually requires less energy for all of the things they do in a day—talking, playing sports, thinking, maintaining body temperature, heart beating—than a single 100W light bulb.

HOW MUCH ENERGY DO WE USE AT HOME:
The average Colorado Household uses about 7,500 kWh electricity and 948 therms of natural gas per year, or about 625 kWh and 79 therms per month. For comparison, the average household electricity usage for the entire U.S. is about 11,000 kWh per year. Space heating and cooling is the number one energy use in the average household, with space heating typically accounting for 50% or more of total household energy use.

Students will be analyzing how much energy their own household uses during the *Personal Energy Audit Design Brief*.

INTRODUCTION TO
THE PERSONAL ENERGY AUDIT DESIGN BRIEF

The *Personal Energy Audit Design Brief* is completed at home by students over the course of one week with students researching and recording detailed data about their home energy use. **See the *Audit* for full details and instructions.**

HAND OUT THE AUDIT:
- Have students put their name on the front cover.
- Assign due dates and have the students fill them into the appropriate spaces.
- Carefully go through the requirements, specifications and restrictions. Students do not write on the grading rubric on the last page. Make sure that everyone understands that deviating from or misinterpreting these requirements will affect their final grade.
- Students should begin collecting data at home immediately.

REVIEW AUDIT PROCEDURES:
- Go over the activity procedures and the 5 sections of the *Audit* (weather log, resource usage, data collection log, results graphs, and activity assessment).
- Review the formulas used in the *Audit*: watts into KWh, converting KWh into BTU's, heating/cooling degree days, BTU's into tons of CO_2.

2.2. APPLIANCE ACTIVITY:

Resources & Materials Needed:
- Access to appliances and tools around the school and/or an assortment of household electrical devices brought into class.
- Kill-a-Watt or other watt meter (optional)
- A copy for each student of the semester reading ("Affluenza" or other book).

Students will need to evaluate the energy use of appliances and other energy-using devices in their homes to complete their Audits. The Energy Scavenger Hunt and/or Watt's Up (you can do one or both activities) will prepare them for this task.

ENERGY SCAVENGER HUNT:
- Show students how to locate and understand the energy rating on appliances and other devices.
- Place students in small teams (2-3).
- Send teams on a scavenger hunt around the school to locate appliance ratings. You can require them to find 5-10 different appliances or give them a list of appliances to find.

Note: Some appliances may only list amps and not watts on their labels. Make sure that students know that Watts are equal to amps multiplied by volts:

Amp x Volts = Watts

For example, in the case of a 7.5 amp toaster connected to a standard 120volt outlet, students can calculate:

7.5 Amps x 120 Volts = 900 Watts

WATT'S UP?:

- Bring from home and/or collect from around school an assortment of common energy-using devices—toaster, lamp, laptop, video game console, hair drier, blender, etc.
- Divide students into small teams and have them compete to guess the watt usage of one item at a time.
- After teams write down their guesses for an item and submit them, read the actual watt usage from the item label plate, or, if you have a watt meter available, plug-in and turn on the item to show its actual use.
- Score each team based on which team gets the closest to the actual answer (the team with the closest answer receives the most points), with the team with the highest final score winning at the end of the game. (Alternately, you can award teams a number of points equal to the number of watts their guess was off by, in which case the team with the LOWEST score wins).

You many want to start out with easier items like light bulbs, and you may also want to give students the hint that (in general) electronics use little energy, devices with motors use more, and items with resistance-heating elements use the most energy.

SEMESTER READING ASSIGNMENT

Teachers may want to introduce a semester book assignment during this unit, and have students read the book throughout the semester, one section at a time. Take a day every couple of weeks to discuss the assigned sections read.

- ***Affluenza,*** by John de Graaf, David Wann, and Thomas Naylor

2.3. CHOICE OF ACTIVITIES

See individual activity guides for the resources & materials needed for each activity.

ADDITIONAL ACTIVITIES:
Some or all of these activities can be done during the course of the unit as additional activities. See individual activity guides for details and materials needed.

SCHOOL BUILDING SURVEY
By the National Energy Education Development Project.
PDF on CD in "Unit 2" folder, available from www.need.org.

This activity involves students conducting a survey of the school premises and recording factors that affect energy use.

ENERGY ACTION PROJECT
From *Energy for Keeps*, by the Energy Education Group.
PDF available for $6 at www.energyforkeeps.org.

This activity has students survey adult attitudes in their own community in order to raise student and public awareness about the use of renewable energy for the generation of electricity. Students research the resources currently used to generate the electricity they use, and determine what renewable resources are available in their region.

LIGHT BULB ACTIVITY
Compact Fluorescent Lamp Demonstration
By Pierce Cedar Creek Institute
PDF on CD in "Unit 2" folder, available from http://www.cedarcreekinstitute.org.
OR
Light Bulbs or Heat Bulbs?
From *Smart Energy Living: Hands-on Activities for the Middle Grades*, by the Colorado Energy Science Center
PDF on CD in "Unit 2" folder, available from www.energyscience.org.

Both of these activities have students study incandescent and compact fluorescent light bulbs, and compare light-output, energy use, and life-cycle costs. This activity is an excellent transition to the Efficiency Unit.

2.4. PERSONAL ENERGY AUDIT COMPLETION AND PRESENTATIONS:

FINISHING THE ENERGY AUDIT:
- Take some time in class to give students a chance to do any final work needed to fully complete their *Audits* and answer any final questions. Insure that they have:
    - o Answered all questions in the *Audit*
    - o Completed all graphs and data collection sheets

STUDENT PRESENTATIONS PREPARATION:
On the final day of Unit 2, students will each give a presentation on their *Audit* results.

- Explain the requirements of and content of the presentations:
    - o Students will present their audit results, including an explanation of their graphs, data collection, etc.
    - o Discuss their carbon footprint (tons of CO_2)
    - o Each presentation should be 4-5 minutes long
- If students have fully completed their *Audits* they can use any remaining class time to work on preparing their presentations.

STUDENT PRESENTATIONS:
- Have students take turns each giving a 4-5 minute presentation on their energy audit findings.
- Collect students' completed *Audits* for grading.

UNIT 3:
Building Energy Efficiency

UNIT OBJECTIVES:
- Increase the students' awareness of what factors affect energy efficiency in homes and how using different materials can affect a home's energy performance.
- Apply knowledge of energy use and efficiency to real life situations.
- Compare different building materials and techniques to achieve the most efficient structure.

VOCABULARY:

Energy	Energy Star	Solar mass
Energy efficiency	Climate	Passive solar
R-Value	Energy transference	Shading
U-Value	BTU	Solar gain

PREREQUISITE KNOWLEDGE:
Students should understand the Laws of Thermodynamics and how this affects building materials. Students should also have an understanding of geographic positioning so the home can make the most of the south facing wall.

ALTERNATIVE ACTIVITY:
Keep It Cool Project

COLORADO MODEL CONTENT STANDARDS: BENCHMARKS FOR GRADES 9-12
1.1 Ask questions and state hypotheses, using prior scientific knowledge to help design and guide development and implementation of a scientific investigation.

1.2 Select and use appropriate technologies to gather, process, and analyze data to report information related to an investigation.

1.3 Identify major sources of error or uncertainty within an investigation. (for example: particular measuring devices and experimental procedures).

1.4 Recognizing and analyzing alternative explanations and models.

1.5 Construct and revise scientific explanations and models, using evidence, logic, and experiments that include identifying and controlling variables.

1.6 Communicate and evaluate scientific thinking that leads to particular conclusions.

2.5. Quantitative relationships involved with thermal energy can be identified, measured, calculated and analyzed (for example: heat transfer in a system involving mass, specific heat, and change in temperature of matter).

2.6. Energy can be transferred through a variety of mechanisms and in any change some energy is lost as heat (for example: conduction, convection, radiation, motion, electricity, chemical bonding changes).

2.8. Quantities that demonstrate conservation of mass and conservation of energy in physical interactions can be measured and calculated.

5.4 There are cause - effect relationships within systems (for example: the effect of temperature on gas volume, effect of CO_2 level on the greenhouse effect, effects of changing in nutrients at the base of a food pyramid).

UNIT 3 TEACHER SCHEDULE

UNIT TIME PERIOD:
This unit is designed to take 18 45-minute class periods.

3.1. WHAT MAKES AN EFFICIENT HOME?
Day 1
Materials: Renewable Energy in Building Design PowerPoint, Copies of NEED Efficiency and Conservation handout

3.2. THE *BUILDING EFFICIENT ARCHITECTURAL MODELS (B.E.A.M.)* DESIGN BRIEF
Day 2: Introduction to the B.E.A.M.
Materials: *Copies of the B.E.A.M. design brief for each student*
Day 3: B.E.A.M. Design Brief, parts 1 and 2
Day 4: Finalizing team designs
Day 5: Construction and safety procedures
Days 6-16: Model construction and preliminary testing
Materials:
- 1/2" plywood bases for the structures to be mounted (20"x20")
- Assorted materials for building model construction--different types of insulation, cardboard, plywood, thin clear plastic sheets for windows, etc.
- Students may bring materials from home if safe for the classroom and this project.
- Assorted methods for fastening--wood glue, hot glue, screws, nails
- Standard tools--hammers, screw drivers, utility knives (safety version, self-retracting), tape measures, *data loggers such as Vernier Lab Pros with temperature probes* (or 12"-18" thermometers).
- computers for web research
Day 17: Model testing
Day 18: Presentations

3.1. WHAT MAKES AN EFFICIENT HOME?

Resources & Materials Needed:
- *Renewable Energy in Building Design* PowerPoint
- Copies of NEED *Efficiency and Conservation* handout for each student

For the *B.E.A.M. Design Brief* student will need an understanding of the factors that affect building efficiency:
- Review what uses energy in a home (from Unit 2)
- Present the *Renewable Energy in Building Design* PowerPoint by Sara Farrar-Nagy.
- Give each student a copy of the NEED *Efficiency and Conservation* handout

HOME EFFICIENCY EVALUATION ASSIGNMENT

Give students a homework assignment to audit the construction and design of their own homes by evaluating the following:
- Window types and efficiency
- Type of construction and materials—siding materials, framing, roofing, etc.
- Levels of insulation in walls and attic
- Energy efficiency of major appliances: Stove, refrigerator, water heater, furnace, air conditioner
- Location and orientation of building and windows
- Building type (apartment, town home, single family home) and style (number of stories, shape, layout)
- Any other factors they think might affect efficiency (landscaping, near by features, roof overhangs, etc)

3.2. THE *BUILDING EFFICIENT ARCHITECTURAL MODELS (B.E.A.M.) DESIGN BRIEF*

INTRODUCTION TO B.E.A.M.

Resources & Materials Needed:
- Copies of the *B.E.A.M.* design brief for each student.

The *B.E.A.M. Design Brief* has students research, design, build, test, and improve a structure to achieve the highest energy efficiency possible. Structures will be tested outside on a sunny day for eight hours with temperature changes being recorded each hour. Students will gain an understanding of how the combination of building location and geographic positional orientation along with building design and materials can greatly affect the energy efficiency of a building. **See the *B.E.A.M.* design brief for full details and instructions.**

HAND OUT THE *B.E.A.M.* DESIGN BRIEF:
- Divide students into groups of two to four.
- Have students put their name on the front cover and all the names of the members of their team on the instructor test data sheet.
- Assign due dates and have the students fill them into the appropriate spaces.
- Carefully go through the requirements, specifications and restrictions. Students do not write on the grading rubric on the last page. Make sure that everyone understands that deviating from or misinterpreting these requirements will affect their final grade.
- Go over the activity procedures and the 7 sections of the brief (research questions, design possibilities and final design, resource usage, data collection log, preliminary testing, activity assessment, instructor's final testing data).

B.E.A.M., PARTS 1 & 2

Have students begin to work on their *B.E.A.M. Design Briefs*:
- Students should begin to answer the research questions in Part 1.
- Students can begin working on their individual designs for Part 2.
- Students should come to the next class with the first page of Part 2 completed (individual Designs 1 & 2).

FINALIZING TEAM DESIGNS

Students meet in their groups to discuss the final design:
- Each student should bring two unique designs to their group.
- Once the team has decided on a final design each member must add this to their design brief.

Check point 1 - No group may progress beyond this point without this step being signed off by the instructor.

CONSTRUCTION & SAFETY PROCEDURES

Have students develop their construction and safety procedures:
- Each student will write their own construction steps and safety procedures for Part 3.
- After each student has created their own steps, have each team decide on a group set of procedures that will be used for the construction of the final structure.
- The team should pick the instruction that they wish to use and have justification for this.

Check point 2 - No group may progress beyond this point without this step being signed off by the instructor.

Team members can now begin organizing materials and tools to begin constructing their structure next class.

MODEL CONSTRUCTION & PRELIMINARY TESTING

Resources & Materials Needed:
- 1/2" plywood bases for the structures to be mounted (20"x20")
- Assorted materials for building model construction--different types of insulation, cardboard, plywood, thin clear plastic sheets for windows, etc.
- Students may bring materials from home: they must be safe for the classroom & this project.
- Assorted methods for fastening--wood glue, hot glue, screws, nails
- Standard tools--hammers, screw drivers, utility knives (safety version, self-retracting), tape measures 12"-18" thermometers (one for each team).
- Computers for web research.

Teams begin the construction phase, completing their design brief as appropriate. Teams should do preliminary testing of their structure and making changes as necessary (Parts 4 & 5).

TESTING PROCEDURE:
- A hole should be placed at the base of the north facing wall (out of direct sunlight) so that the reading will be taken at floor level inside the structure.
- The hole can be sealed by the team with a material that won't permanently affect the thermometer.
- The thermometer must be able to be read from the outside of the structure.
- All structures should start at room temperature inside the classroom and a beginning reading taken then, before going outside.
- Students should also measure ambient air temperature
- Optionally, students can measure incident solar energy (w/m^2) using a solar meter or pyronometer (Dodge Products model # 776E)

ALTERNATIVE TESTING:
If you are limited to working inside, you can place a hand warmer or cold pack inside the structure to simulate heating or cooling to test efficiency in an otherwise constant temperature environment. Do not allow the thermometer to come into contact with the heating or cooling device as this will skew the results.

On the final day of preliminary testing make sure all students are prepared for the final test of their models.

FINAL MODEL TESTING

Students will test their models over the course of a school day:
- Teams meet before first hour to set up their structure.
- One member takes readings each hour for eight hours (8am-3pm). Student must get the official record sheet for their team from the instructor each hour.
- Teams collect their structure after the last class of the day and record final results.
- During the regular class period students should be completing unfinished sections of the design brief.
- Design briefs are due next class period, and each team will give a 3-5 minute presentation on their structure and the results of their testing.

PRESENTATIONS

Teams each give a 3-5 minute presentation on their structure, design process, and testing results.
Students turn in their completed design briefs.

UNIT 4:
Solar Energy

UNIT OBJECTIVES:
*   Increase the students awareness of different types of solar energy capture.
*   Students should be able to discuss what factors enhance or hinder their attempts to gather the sun's energy.

VOCABULARY:

Entropy	Parabolic trough	Solar spectrum
Solar thermal	BTU	Active solar
Solar cell	Photon	Solar gain
photovoltaic	multimeter	Inverter

COLORADO MODEL CONTENT STANDARDS: BENCHMARKS FOR GRADES 9-12

1.1 Ask questions and state hypotheses, using prior scientific knowledge to help design and guide development and implementation of a scientific investigation.

1.2 Select and use appropriate technologies to gather, process, and analyze data to report information related to an investigation.

1.3 Identify major sources of error or uncertainty within an investigation (for example: particular measuring devices and experimental procedures).

1.4 Recognizing and analyzing alternative explanations and models.

1.5 Construct and revise scientific explanations and models, using evidence, logic, and experiments that include identifying and controlling variables.

1.6 Communicate and evaluate scientific thinking that leads to particular conclusions.

2.4. Word and chemical equations are used to relate observed changes in matter to its composition and structure (for example: conservation of matter).

2.5. Quantitative relationships involved with thermal energy can be identified, measured, calculated and analyzed (for example: heat transfer in a system involving mass, specific heat, and change in temperature of matter).

2.6. Energy can be transferred through a variety of mechanisms and in any change some energy is lost as heat (for example: conduction, convection, radiation, motion, electricity, chemical bonding changes).

2.7. Light and sound waves have distinct properties; frequency, wavelengths and amplitude.

2.8. Quantities that demonstrate conservation of mass and conservation of energy in physical interactions can be measured and calculated.

4.4. There are costs, benefits, and consequences of natural resource exploration, development, and consumption (*for example: geosphere, biosphere, hydrosphere, atmosphere and greenhouse gas*).

4.5.There are consequences for the use of renewable and nonrenewable resources.

4.11. There are factors that may influence weather patterns and climate and their effects within ecosystems *(for example: elevation, proximity to oceans, prevailing winds, fossil fuel burning, volcanic eruptions).*

4.15. There is electromagnetic radiation produced by the Sun and other stars *(for example: X- ray, ultraviolet, visible light, infrared, radio).*

5.1 Print and visual media can be evaluated for scientific evidence, bias and opinion.

5.3 Graphs, equations, or other models are used to analyze systems involving change and constancy (for example, comparing the geologic time scale to shorter time frames, exponential growth, a mathematical expression for gas behavior; constructing a closed ecosystem such as an aquarium).

5.4 There are cause - effect relationships within systems (for example: the effect of temperature on gas volume, effect of CO_2 level on the greenhouse effect, effects of changing in nutrients at the base of a food pyramid).

5.6 Interrelationships among science, technology and human activity lead to further discoveries that impact the world in positive and negative ways.

UNIT TIME PERIOD:
This unit is designed to take 26 45-minute class periods.

4.1 SOLAR THERMAL

Day 1: Basics of solar energy and introduction to *BTU or Bust Design Brief*
Materials: *Copies of the BTU or Bust Design Brief for each student, copies of the NEED Solar handout for each student.*
Day 2: Introduction to solar energy technologies and *BTU or Bust* parts 1 and 2
Materials: *Copies of solar technology handouts for each student: History of the Solar Cell, Passive Solar Design, Parabolic Trough Water Heating, Solar Hot Water Heating Systems.*
Day 3: Finalize team designs
Day 4: Water heater construction safety procedures
Days 5-15: Water heater construction and preliminary testing
Materials:

- *An 18" piece of 1-1/2" PVC pipe with a 45 degree angle at one end (thermometer end), capped at both ends (a hole needs to be drilled in the cap on the 45 degree angle so that the thermometer can be inserted to record temperatures)*
- *A variety of reflective materials, cardboard, wood scraps, etc.*
- *Safety Glasses*
- *Assorted methods for fastening--wood glue, hot glue, screws, nails.*
- *Typical shop tools--hammers, screw drivers, utility knives (safety version, self-retracting), tape measures*
- *12"-18" thermometers (one for each team), for final testing use data logger such as Vernier Lab Pro with temperature probe instead of thermometers if available.*
- *Multimeters*

Day 16: Water heater final testing
Day 17: Presentations of designs and testing results

4.2 SOLAR PHOTOVOLTAICS

Day 18: Introduction to Photovoltaic Solar Cells
Materials: *Potential of Photovoltaics PowerPoint*
Days 19-20: *The Power of the Sun* video
Materials: *The Power of the Sun DVD (available at http://powerofthesun.ucsb.edu/)*
Days 21-25: Solar Photovoltaics Project
Materials: *See the NEED Photovoltaics Project teacher's guide for materials list.*
Day 26: Sizing and installing solar energy systems

4.1. SOLAR THERMAL

BASICS OF SOLAR ENERGY

Directly or indirectly, virtually all energy we use is derived from solar energy: wind is created when solar energy creates differences of temperature on areas of the earth's surface, biomass energy is derived from solar energy that has been captured and stored, and fossil fuels are derived from biomass that has been trapped and transformed by geologic processes over millions of years.

In addition to the many indirect ways we use the sun's energy, there are three main ways that we use solar energy directly:

Passive solar: Passive solar is when buildings directly take advantage of solar energy. A green house is an example of a passive-solar structure. While virtually all structures capture passive solar energy to some extent, sophisticated passive solar buildings are precisely designed (using window placement, building orientation and shape, thermal

mass, etc) to take maximum advantage of solar energy during the course of the year. A trombe wall is an example of passive solar technology: This is when a wall (often dark-colored) with large amounts of thermal mass is placed between large south-facing windows and the main building. The thermal mass of the wall captures large amounts of solar energy during the day and then slowly releases that heat into the building during the night, resulting in a more consistent internal temperature.

A trombe wall

Solar Thermal: Solar thermal systems actively collect, transport, and utilize solar energy to generate heat. The most common systems are those used to heat water, but there are also systems designed for other applications like space heating and cooking. The *BTU or Bust Design Brief* has students design and build a solar thermal water heater.

Solar photovoltaic: Solar electric systems generate electricity using sunlight. Solar electric technology will be covered in detail in the second half of this unit (following the *BTU or Bust Design Brief*).

- Hand out copies of the NEED *Solar* energy fact sheet.

INTRODUCTION TO *BTU OR BUST DESIGN BRIEF*

The *Btu or Bust Design Brief* has students design, build and test a parabolic trough that will heat one pound of water to as high of a temperature as can be achieved without damaging the water storage container. Experimenting with different types of materials will also allow them to understand how the properties of different materials can drastically affect the outcome of their experiment. **See the design brief for full details and instructions.**

HAND OUT THE DESIGN BRIEF:

- Break students into groups of two to four. *Alternative: Sometimes it works better to wait until they have their individual design ideas before you break them into groups.*
  - Have students put their name on the front cover and all the names of the members of their team on the instructor test data sheet.
  - Assign due dates and have the students fill them into the appropriate spaces.
  - Carefully go through the requirements, specifications and restrictions. Students do not write on the grading rubric on the last page. Make sure that everyone understands that deviating from or misinterpreting these requirements will affect their final grade.

REVIEW THE DESIGN BRIEF PROCEDURES:

- Go over the activity procedures and the sections of the design brief (research, design, resource usage, design data log, preliminary testing, final test data, and activity assessment).
- Review the rules for the design brief and the schedule and guidelines for design, testing, and final evaluation.

INTRODUCTION TO SOLAR ENERGY TECHNOLOGIES

Resources & Materials Needed:
- Copies of solar technology handouts for each student: *History of the Solar Cell, Passive Solar Design, Parabolic Trough Water Heating, Solar Hot Water Heating Systems*

Give students copies of the following solar technology handouts:
- History of the Solar Cell
- Passive Solar Design
- Parabolic Trough Water Heating
- Solar Hot Water Heating Systems

BTU OR BUST PARTS 1 AND 2

Students begin working on their *BTU or Bust Design Briefs:*
- Students should begin to answer the research questions in Part 1.
- Students can begin working on their individual designs for Part 2.
- Students should come to the next class with the first page of Part 2 completed. (Design 1 & 2)

FINALIZE TEAM DESIGNS

STUDENTS FINALIZE TEAM DESIGNS:
- Students meet in their groups to discuss the final design. Each student should bring two unique designs to the table. *Alternative: Now break them into groups of 2 – 4.*
- Once the team has decided on a final design each member must add this to their design brief.

Check point 1 - No group may progress beyond this point without this step being signed off by the instructor.

WATER HEATER CONSTRUCTION SAFETY PROCEDURES

Teams develop safety procedures for their projects:
- Each student will then begin writing construction steps and safety procedures for Part 3. Each student will create their own steps and the team will decide which ones to use for the construction of the final structure.
- The team should pick the instruction that they wish to use.

Check point 2 - No group may progress beyond this point without this step being signed off by the instructor.

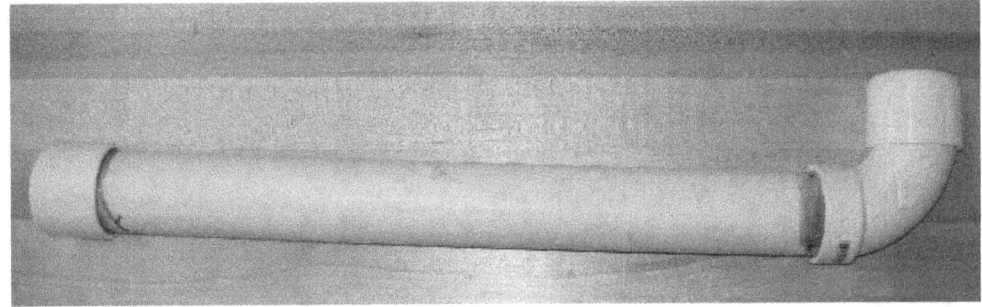
An example of the basic PVC-pipe units used for the water heaters

WATER HEATER CONSTRUCTION AND PRELIMINARY TESTING

Resources & Materials Needed:
- An 18" piece of 1-1/2" PVC pipe with a 45 degree angle at one end (thermometer end), capped at both ends (a hole needs to be drilled in the cap on the 45 degree angle so that the thermometer can be inserted to record temperatures)
- A variety of reflective materials, cardboard, wood scraps, etc.
- Safety Glasses
- Assorted methods for fastening--wood glue, hot glue, screws, nails.
- Typical shop tools--hammers, screw drivers, utility knives (safety version, self-retracting), tape measures
- 12"-18" thermometers (one for each team), for final testing use data loggers such as Vernier Lab Pros with temperature probes instead of thermometers if available.
- Multimeters

Teams begin the construction phase, completing their design brief as appropriate.

Teams begin testing their structure and making changes as necessary, and complete parts 5 and 6 of their design briefs.

TESTING PROCEDURE:
- The hole in the PVC tube can be sealed by the team with a material that won't permanently affect the thermometer or the tube.
- The thermometer must be able to be read from the outside of the device.
- All devices should start at tap temperature inside the classroom and an initially recorded temperature, before going outside.

WATER HEATER FINAL TESTING

Students will test their water heaters over the course of a full school day:
- Teams meet before first hour to set up their device.
- One member takes readings each hour for eight hours (~8:00-3:00). Student must get record sheet from instructor each hour. If the school has electronic data recorders, the students can program them to take readings automatically.
- Teams collect their device after the last class of the day and record final results.
- During the regular class time students should work on completing unfinished sections of the design brief.
- Completed design briefs will be due next class period.
- Each team will give a 3-5 minute presentation on their device and the results of their testing.

PRESENTATIONS OF DESIGNS & TESTING RESULTS

Each team gives a 3-5 minute presentation of their device and results.
Students turn in their completed design briefs.

4.2. SOLAR PHOTOVOLTAICS

INTRODUCTION TO PHOTOVOLTAIC SOLAR CELLS

Resources & Materials Needed:
• *Potential of Photovoltaics* PowerPoint

Present to students *The Potential of Photovoltaics* PowerPoint, by Brent Nelson of NREL. The presentation covers:

• A basic introduction to photovoltaic technology.
• The potential for solar energy development in the context of global energy use.
• The current status of photovoltaic technologies.
• Challenges and issues related to solar energy development.

For additional information on solar energy research and development, see the National Renewable energy Laboratory website: http://www.nrel.gov/solar/

You may want to consider arranging a class room trip and presentation with the NREL visitor center in Golden. See the NREL website for details on how to arrange a program.

THE POWER OF THE SUN

Show students *The Power of the Sun* (or an alternate video on solar energy). "The Power of the Sun" consists of two films on a single DVD, described by the films' website:

> "The Power of the Sun-The Science of the Silicon Solar Cell" (S), is a 20-minute animated educational film for 12th grade High School students, or freshman College/University students with interests in physics and/or chemistry, materials science, engineering. (The silicon solar cell is currently the most important generator of solar electricity).
>
> "The Power of the Sun" (G) is a 56-minute film, telling the story of photovoltaics -Light; History and Science; Implementation; and Future. It is designed for general public with interest in science, its history and its current and future applications to the world's energy needs, as well as for policy-makers and opinion leaders in the field of energy. It is also highly recommended for students who are using the 20-minute science film (S), to provide them with a broad perspective."

For more information, or to order the DVD ($10, or $5 for teachers), visit:

http://powerofthesun.ucsb.edu/

On the website there is also an online presentation version (under "instructional materials") of "The Science of the Silicon Solar Cell."

SOLAR PHOTOVOLTAICS PROJECT

Spend five days doing the NEED Solar Photovoltaics project. This project requires NEED's solar cell kit designed for this project (available from NEED for $350). See the teachers guide PDF for full materials list (page 3) and complete details and instructions (page 7).

ALTERNATE ACTIVITIES:
See the NREL *Solar Energy Science Projects* for alternatives to the NEED photovoltaic activity.

SIZING AND INSTALLING SOLAR ENERGY SYSTEMS

Introduce students to the steps involved in real-world process of designing and installing solar electric systems. Typically the process involves:

- Deciding on the type of system to be installed
  - Off grid, Grid inter-tied, grid inter-tied with battery back-up
- Estimating current electricity use
- Estimating the available solar resource at the site
  - Available solar energy
  - Orientation options, possible obstructions
- Sizing installation
  - Available area
  - Desired energy production
- Select equipment components
  - PV Modules
  - Inverter
  - Mountings or racks
  - Battery system (for off-grid or back-up)
  - Additional components (wiring, disconnects, etc)
- Economics
  - Equipment and installation costs
  - Rebates and incentives
  - Financing, Pay-back period, return on investment
- Permitting & Installation
- Monitoring & Maintenance

The "My Solar Estimator" tool on the website http://www.findsolar.com/ will generate a simplified estimate of the size, cost, and energy production of a theoretical systems based on your CO County, utility, and energy demand.

You may want to have students to use this site (or similar online calculator) to develop an estimate of a solar energy system for their own homes.

On this day you may want to invite a guest speaker from a local solar energy installation company to talk about their experience designing and installing solar energy systems.

UNIT 5:

Wind Energy

UNIT OBJECTIVES:
- Increase the students awareness of the potential of wind energy.
- Apply knowledge of calculating swept area to wind turbine blade design.
- Build and test wind turbine blades based on knowledge gained.

VOCABULARY:

Wind turbine	Nacelle	Yaw
Blade	Generator	Aerodynamic lift
Swept area	Pitch	Anemometer

COLORADO MODEL CONTENT STANDARDS: BENCHMARKS FOR GRADES 9-12

1.1 Ask questions and state hypotheses, using prior scientific knowledge to help design and guide development and implementation of a scientific investigation.

1.2 Select and use appropriate technologies to gather, process, and analyze data to report information related to an investigation.

1.3 Identify major sources of error or uncertainty within an investigation. (for example: particular measuring devices and experimental procedures.

1.4 Recognizing and analyzing alternative explanations and models.

1.5 Construct and revise scientific explanations and models, using evidence, logic, and experiments that include identifying and controlling variables.

1.6 Communicate and evaluate scientific thinking that leads to particular conclusions.

2.6. Energy can be transferred through a variety of mechanisms and in any change some energy is lost as heat (for example: conduction, convection, radiation, motion, electricity, chemical bonding changes).

2.8. Quantities that demonstrate conservation of mass and conservation of energy in physical interactions can be measured and calculated.

4.4. There are costs, benefits, and consequences of natural resource exploration, development, and consumption (*for example: geosphere, biosphere, hydrosphere, atmosphere and greenhouse gas*).

4.5.There are consequences for the use of renewable and nonrenewable resources.

4.8. Energy transferred within the atmosphere influences weather *(for example: the role of conduction, radiation, convection, and heat of condensation in clouds, precipitation, winds, storms).*

4.9.   Weather is caused by differential heating, the spin of the Earth and changes in humidity (air pressure, wind patterns, coriolis effect).

4.11. There are factors that may influence weather patterns and climate and their effects within ecosystems *(for example: elevation, proximity to oceans, prevailing winds, fossil fuel burning, volcanic eruptions).*

5.1 Print and visual media can be evaluated for scientific evidence, bias and opinion.

5.2 Identify reasons why consensus and peer review are essential to the Scientific Process.

5.3 Graphs, equations, or other models are used to analyze systems involving change and constancy (for example, comparing the geologic time scale to shorter time frames, exponential growth, a mathematical expression for gas behavior; constructing a closed ecosystem such as an aquarium).

5.4 There are cause - effect relationships within systems (for example: the effect of temperature on gas volume, effect of CO_2 level on the greenhouse effect, effects of changing in nutrients at the base of a food pyramid).

5.6 Interrelationships among science, technology and human activity lead to further discoveries that impact the world in positive and negative ways.

UNIT 5 TEACHER SCHEDULE

UNIT TIME PERIOD:
This unit is designed to take 12 45-minute class periods.

5.1. WIND ENERGY AND THE WIND TURBINE BLADE DESIGN BRIEF
Day 1: Introduction to wind energy and the *Wind Turbine Blade Design Brief*
Materials:
- *Wind Turbine Blade Design Briefs*
- *Basic Wind Background PowerPoint*
- *Windmill Blade Design Challenge handouts*
- *Windmill Design Guide handouts*
- *Wind Energy Teachers Guide*
- *Renewable Energy Source: Wind, from Energy for Keep handouts.*

Day 2: Wind resource and technology and *Turbine Blade Design Brief* parts 1 and 2
Materials: *Wind Resource and Technology PowerPoint*
Day 3: Finalizing team turbine blade designs
Day 4: Blade construction steps and safety procedures
Days 5-10: Blade construction and preliminary testing
Materials: *KidWind turbine kits, Box fans, standard shop tools, Anemometer (optional)*
Day 11: Final blade testing
Day 12: Turbine blade project presentations

5.1. WIND ENERGY &
THE *WIND TURBINE BLADE DESIGN BRIEF*

INTRODUCTION TO WIND ENERGY

> **Resources & Materials Needed:**
> * *Wind Turbine Blade Design Briefs*
> * *Basic Wind Background* PowerPoint
> * *Windmill Blade Design Challenge* handouts
> * *Windmill Design Guide* handouts
> * *Wind Energy Teachers Guide*
> * *Renewable Energy Source: Wind,* from *Energy for Keeps* handouts.

Present the *Basic Wind Background* (by KidWind) PowerPoint to students. The presentation introduces:
* The types of wind energy technologies
* Existing wind energy development
* The future potential for wind energy
* Basic wind energy design principles.

The American Wind Energy Association's *Wind Energy Teachers Guide* provides useful background information to help you prepare your presentation.

Give each student a copy and assign as reading homework the wind energy and turbine blade design hand outs:
* *Windmill Blade Design Challenge,* KidWind Project, (www.kidwind.org).
* *Windmill Design Guide,* by Hugh Piggott (www.scoraigwind.com)
* *Renewable Energy Source: Wind,* From *Energy for Keeps*, by the Energy Education Group.

INTRODUCTION TO
THE WIND TURBINE BLADE DESIGN BRIEF

The *Wind Turbine Blade Design Brief* has students design, build and test a wind turbine blade to achieve the possible electrical output. This project is meant to use wind energy kits available from KidWind (for kit options and prices, see www.kidwind.org). **See the design brief for full details and instructions.**

HAND OUT THE DESIGN BRIEF:
- Break students into groups of two to four. *Alternative: Sometimes it works better to wait until they have their individual design ideas before you break them into groups.*
    - Have students put their name on the front cover and all the names of the members of their team on the instructor test data sheet.
    - Assign due dates and have the students fill them into the appropriate spaces.
    - Carefully go through the requirements, specifications and restrictions. Students do not write on the grading rubric on the last page. Make sure that everyone understands that deviating from or misinterpreting these requirements will affect their final grade.

REVIEW THE DESIGN BRIEF PROCEDURES:
- Go over the activity procedures and the sections of the design brief.
- Review the rules for the design brief and the schedule and guidelines for design, testing, and final evaluation.

WIND RESOURCE AND TECHNOLOGY

Resources & Materials Needed:
- *Wind Resource and Technology* PowerPoint

Present the *Wind Resource and Technology* PowerPoint by Robi Robichaud, NREL. The presentation provides more advanced technical information and data on the factors involved in wind energy systems.

BLADE DESIGN BRIEF
PARTS 1 AND 2

Students begin working on their turbine blade design briefs:
- Students should then begin to answer the research questions in Part 1.
- Students can begin working on their individual designs for Part 2.
- Students should come to the next class with the first page of Part 2 completed. (Design 1 & 2)

FINALIZING TEAM BLADE DESIGNS

Students finalize team designs:
- Students meet in their groups to discuss the final design. Each student should bring two unique designs to the table. *Alternative: Now break them into groups of 2 - 4*
- Once the team has decided on a final design each member must add this to their design brief.

Check point 1 - No group may progress beyond this point without this step being signed off by the instructor.

BLADE CONSTRUCTION STEPS & SAFETY PROCEDURES

Teams develop safety procedures for their projects:
- Each student will then begin writing construction steps and safety procedures for Part 3. Each student will create their own steps and the team will decide which ones to use for the construction of the final structure.
- The team should pick the instruction that they wish to use.

Check point 2 - No group may progress beyond this point without this step being signed off by the instructor.

Teams can begin organizing materials and tools to begin constructing their device next class.

BLADE CONSTRUCTION & PRELIMINARY TESTING

Resources & Materials Needed:
- KidWind turbine kits
- Box fans
- Anemometer (optional)

Teams begin the construction phase, completing their design brief as appropriate.

Teams begin testing their blades and making changes as necessary (Parts 4 & 5 of the brief).

TESTING PROCEDURE:
- Using the Kid Wind Turbines kits, multimeters, and box fans, teams test the power output of their blade designs at various wind speeds.
- Make sure that teams understand what the final testing conditions (fan distance and sped settings, etc) will be so they can design accordingly. If an anemometer is available then have students determine and consider the actual recorded wind speed.

FINAL BLADE TESTING

- Team do final testing of the wind turbine blades
- Students complete any remaining sections of their design briefs, which will be due next class period.
- Have each team prepare a 3-5 minute presentation on their device and the results of their testing for day twelve.

TURBINE BLADE PROJECT PRESENTATIONS

Teams give 3-5 minute presentations on their device and results.

Students turn in their completed design briefs.

Spring Semester

UNIT 6:
Hydrogen and Fuel Cells

UNIT OBJECTIVES:
- Increase the student awareness of different types of fuel cells, how they work and fuels they use.
- Apply knowledge of fuel cells to real life situations.
- Compare different building materials and techniques to achieve the most efficient outcome.
- Students should be able to discuss different types of fuel cells, fuels and how they work as well as how their application to both stationary and transportation situations can be beneficial.

VOCABULARY:
P.E.M.	Fuel cell	Reformation
Electrolysis	Electrolyte	Electrolyzer
Hydrogen	Energy carrier	

ALTERNATIVE ACTIVITIES:
- College Fuel Cell Class. See activity guide for details and instructions.

COLORADO MODEL CONTENT STANDARDS: BENCHMARKS FOR GRADES 9-12
1.1 Ask questions and state hypotheses, using prior scientific knowledge to help design and guide development and implementation of a scientific investigation.

1.2 Select and use appropriate technologies to gather, process, and analyze data to report information related to an investigation.

1.3 Identify major sources of error or uncertainty within an investigation (for example: particular measuring devices and experimental procedures).

1.4 Recognizing and analyzing alternative explanations and models.

1.5 Construct and revise scientific explanations and models, using evidence, logic, and experiments that include identifying and controlling variables.

1.6 Communicate and evaluate scientific thinking that leads to particular conclusions.

2.4. Word and chemical equations are used to relate observed changes in matter to its composition and structure (for example: conservation of matter).

2.5. Quantitative relationships involved with thermal energy can be identified, measured, calculated and analyzed (for example: heat transfer in a system involving mass, specific heat, and change in temperature of matter).

2.6. Energy can be transferred through a variety of mechanisms and in any change some energy is lost as heat (for example: conduction, convection, radiation, motion, electricity, chemical bonding changes).

2.8. Quantities that demonstrate conservation of mass and conservation of energy in physical interactions can be measured and calculated.

5.1 Print and visual media can be evaluated for scientific evidence, bias and opinion.

5.2 Identify reasons why consensus and peer review are essential to the Scientific Process.

5.3 Graphs, equations, or other models are used to analyze systems involving change and constancy (for example, comparing the geologic time scale to shorter time frames, exponential growth, a mathematical expression for gas behavior; constructing a closed ecosystem such as an aquarium).

5.4 There are cause - effect relationships within systems (for example: the effect of temperature on gas volume, effect of CO_2 level on the greenhouse effect, effects of changing in nutrients at the base of a food pyramid).

5.6 Interrelationships among science, technology and human activity lead to further discoveries that impact the world in positive and negative ways.

UNIT TIME PERIOD:
This unit is designed to take 28 45-minute class periods.

6.1. INTRODUCTION TO HYDROGEN
Day 1
Materials: Hydrogen and Fuel Cells PowerPoint, NEED Hydrogen handout
Day 2: Hydrogen fuel cell technologies and history
Materials: Hydrogen Fuel Cells and Fuel Cells: History, Types, and Uses PowerPoints

6.2. "BUILD YOUR OWN FUEL CELL" PROJECT
Days 3-10
Materials:
- Build Your Own Fuel Cell project guide by Phillip Hurley--See the activity guide for full materials list, PDF available for $14.95 at goodideacreative.com.
- Fuel cell kits (can be purchased from www.fuelcellstore.com).

6.3. FUEL CELL INVENTION DESIGN BRIEF
Day 11: Introduction to the *Fuel Cell Invention Design Brief*
Materials: Fuel Cell Invention Design Brief, Guide to Fuel Cells handout.
Day 12: *Fuel Cell Invention Design Brief* parts 1 and 2
Day 13: Finalizing team designs
Days 14-19: Construction, testing, and presentation preparation
Day 20: Project presentations

6.1. INTRODUCTION TO HYDROGEN

Resources & Materials Needed:
- *Hydrogen and Fuel Cells* PowerPoint
- NEED Hydrogen handout

- Present the *Hydrogen and Fuel Cells* PowerPoint, by the DOE. The presentation provides a basic introduction to Hydrogen technology.
- Give each student a copy of the NEED Hydrogen handout, and have them read it and discuss it and the presentation.

HYDROGEN FUEL CELL TECHNOLOGIES & HISTORY

Resources & Materials Needed:
- *Hydrogen Fuel Cells* and *Fuel Cells: History, Types, and Uses* PowerPoints

Present the PowerPoints *Hydrogen Fuel Cells*, by John Turner, NREL and *Fuel Cells: History, Types, and Uses*, by Matthew Brown. These presentations provide more detailed looks at hydrogen technology and applications, covering topics including:
- The history of fuel cell development
- Hydrogen production methods and technology
- Hydrogen distribution
- The details of vehicle and stationary fuel cell types and construction

6.2. BUILD YOUR OWN FUEL CELL PROJECT

Resources & Materials Needed:
- *Build Your Own Fuel Cell* project guide
- See the activity guide for full materials list
- Fuel cell kits (can be purchased from www.fuelcellstore.com).

BUILD YOUR OWN FUEL CELL PROJECT
by Phillip Hurley, PDF available for $14.95 at goodideacreative.com.

This project has students experiment with Proton Exchange Membrane (P.E.M.) fuel cells and electrolyzers, allowing them to produce their own hydrogen using solar energy. The project has students build and test their own fuel cell, using materials that can be purchased at www.fuelcellstore.com.
See the activity guide for full details, materials, and instructions.

6.3. FUEL CELL INVENTION DESIGN BRIEF

INTRODUCTION TO
THE FUEL CELL INVENTION DESIGN BRIEF

Resources & Materials Needed:
- *Fuel Cell Invention* design briefs.
- *Guide to Fuel Cells* handout.

HAND OUT THE DESIGN BRIEF:
- Break students into groups of two to four. *Alternative: Sometimes it works better to wait until they have their individual design ideas before you break them into groups.*
- Have students put their name on the front cover and all the names of the members of their team on the instructor test data sheet.
- Assign due dates and have the students fill them into the appropriate spaces.
- Carefully go through the requirements, specifications and restrictions. Students do not write on the grading rubric on the last page. Make sure that everyone understands that deviating from or misinterpreting these requirements will affect their final grade.

REVIEW THE DESIGN BRIEF PROCEDURES:
- Go over the activity procedures and the sections of the design brief.
- Review the rules for the design brief and the schedule and guidelines for design, testing, and final evaluation.

Hand out and begin reading and discussing the *Guide to Fuel Cells,* by Los Alamos National Laboratory.

DESIGN BRIEF PARTS 1 & 2

Students begin working on their *Fuel Cell Invention Design Briefs*:
- Students should begin to answer the research questions in Part 1.
- Students can begin working on their individual designs for Part 2.
- Students should come to the next class with the first page of Part 2 completed. (Design 1 & 2)

FINALIZING TEAM DESIGNS

STUDENTS FINALIZE TEAM DESIGNS:
- Students meet in their groups to discuss the final design. Each student should bring two unique designs to the table. *Alternative: Now break them into groups of 2 – 4.*
- Once the team has decided on a final design each member must add this to their design brief.

Check point 1 - No group may progress beyond this point without this step being signed off by the instructor.

CONSTRUCTION & PRELIMINARY TESTING

Resources & Materials Needed:
- 0.5 watt PEM fuel cells available at fuelcellstore.com
- Any assorted materials necessary for student inventions.

Have teams begin constructing and testing their inventions:
- Teams can bring their own materials to use if approved by the instructor.
- Teams complete their design brief as appropriate.
- Teams test their inventions and make changes as necessary.

INVENTION PROJECT WRAP UP

Have students prepare to present their inventions:
- Have each team prepare a 3-5 minute presentation on their projects.
- Teams also need to develop a professional looking product brochure about the invention.

PROJECT PRESENTATIONS

- Teams present their inventions and product brochures.
- Students turn in the completed design briefs.

UNIT 7:
Sustainable Transportation

UNIT OBJECTIVES:
- Increase student awareness of different modes of transportation.
- Assess the alternatives to conventional modes of transportation.
- Increase student awareness of future technologies and what it will take to implement those technologies.
- Students should be able to discuss conventional modes of transportation and compare them to sustainable modes of transportation.

VOCABULARY:
Electric car	Transportation	Biofuels
FCV		

COLORADO MODEL CONTENT STANDARDS: BENCHMARKS FOR GRADES 9-12
1.4 Recognizing and analyzing alternative explanations and models.

1.6 Communicate and evaluate scientific thinking that leads to particular conclusions

4.4. There are costs, benefits, and consequences of natural resource exploration, development, and consumption (*for example: geosphere, biosphere, hydrosphere, atmosphere and greenhouse gas*)

4.5.There are consequences for the use of renewable and nonrenewable resources

5.1 Print and visual media can be evaluated for scientific evidence, bias and opinion

5.2 Identify reasons why consensus and peer review are essential to the Scientific Process

5.3 Graphs, equations, or other models are used to analyze systems involving change and constancy (for example, comparing the geologic time scale to shorter time frames, exponential growth, a mathematical expression for gas behavior; constructing a closed ecosystem such as an aquarium)

5.4 There are cause - effect relationships within systems (for example: the effect of temperature on gas volume, effect of CO_2 level on the greenhouse effect, effects of changing in nutrients at the base of a food pyramid)

5.5 Scientific knowledge changes and accumulates over time; usually the changes that take place are small modifications of prior knowledge about major shifts in the scientific view of how the world works do occur

5.6 Interrelationships among science, technology and human activity lead to further discoveries that impact the world in positive and negative ways.

UNIT 7 TEACHER SCHEDULE

UNIT TIME PERIOD:
This unit is designed to take 2 45-minute class periods.

7.1 Introduction to sustainable transportation
Day 1
Materials: *Northeast Sustainable Energy Association transportation handouts*
Day 2: Student transportation presentations

7.2. FUEL CELL VEHICLE DESIGN BRIEF
Day 3: Introduction to the *Fuel Cell Vehicle Design Brief*
Materials: *Fuel Cell Vehicle Design Brief, Guide to Fuel Cells handout.*
Day 4: *Fuel Cell Vehicle Design Brief* parts 1 and 2
Day 5: Finalizing team designs
Day 6: Vehicle construction steps and safety procedures
Days 7-17: Vehicle construction and preliminary testing
Materials: *0.5 watt PEM fuel cells (available at fuelcellstore.com), gear systems, assorted wheels, assorted pieces of wood, cardboard, etc for vehicle construction.*
Day 18: Final testing: Race Day
Day 19: Vehicle project wrap up
Day 20: Project presentations

7.1. INTRODUCTION TO SUSTAINABLE TRANSPORTATION

Resources & Materials Needed:
- Northeast Sustainable Energy Association transportation handouts

ALTERNATIVE TRANSPORTATION TECHNOLOGIES:

Have students brainstorm all the various alternative fuels and modes of transportation to conventional fossil-fuel vehicles, and develop a list that might include:

Alternative Modes:
- Bus
- Bicycle
- Light (electric) rail
- Train
- Walking

Fuels:
- Compressed natural gas (CNG)
- Ethanol
- Biodiesel
- Hydrogen
- Electric
- Hybrid and/or plug-in hybrid

STUDENT TRANSPORTATION PROJECTS:

- Divide students into teams
- Assign or have teams select a transportation mode or alternative fuel from the list (you may want to remove hydrogen and biodiesel as options, since these fuels will be studied in detail later).
- Hand out factsheets to teams on their technology.
- Each team will prepare (in class and as homework) a short presentation for next class on their technology. Each presentation should include:
  - A basic introduction to the technology and how it works
  - The advantages, limitations, and possibilities for the technology.
  - What infrastructure and development is needed to promote the technology (new fueling stations, new fuel distribution systems, new rail lines, walkable/bikeable community design, etc)
  - Examples of where this technology is being promoted and implemented
  - The current uses, availability, and potential for the technology in the students' local area.
- Teams can divide the research for the sections of their presentations amongst team members.

TRANSPORTATION PRESENTATIONS

Student teams give their transportation presentations.

7.2. FUEL CELL VEHICLE DESIGN BRIEF

INTRODUCTION TO
THE FUEL CELL VEHICLE DESIGN BRIEF

Resources & Materials Needed:
* *Fuel Cell Vehicle* design briefs.

HAND OUT THE DESIGN BRIEF:
* Break students into groups of two to four. *Alternative: Sometimes it works better to wait until they have their individual design ideas before you break them into groups.*
* Have students put their name on the front cover and all the names of the members of their team on the instructor test data sheet.
* Assign due dates and have the students fill them into the appropriate spaces.
* Carefully go through the requirements, specifications and restrictions. Students do not write on the grading rubric on the last page. Make sure that everyone understands that deviating from or misinterpreting these requirements will affect their final grade.

REVIEW THE DESIGN BRIEF PROCEDURES:
* Go over the activity procedures and the sections of the design brief.
* Review the rules for the design brief and the schedule and guidelines for design, testing, and final evaluation.

FUEL CELL VEHICLE DESIGN BRIEF PARTS 1 & 2

Students begin working on their *Fuel Cell Vehicle Design Briefs*:
* Students should begin to answer the research questions in Part 1.
* Students can begin working on their individual designs for Part 2.
* Students should come to the next class with the first page of Part 2 completed. (Design 1 & 2)

Honda's hydrogen fuel cell FCX Clarity.

FINALIZING TEAM DESIGNS

STUDENTS FINALIZE TEAM DESIGNS:
- Students meet in their groups to discuss the final design. Each student should bring two unique designs to the table. *Alternative: Now break them into groups of 2 – 4.*
- Once the team has decided on a final design each member must add this to their design brief.

Check point 1 - No group may progress beyond this point without this step being signed off by the instructor.

VEHICLE CONSTRUCTION STEPS
& SAFETY PROCEDURES

Have students develop their construction and safety procedures:
- Each student will write their own construction steps and safety procedures for Part 3.
- After each student has created their own steps, have each team decide on a group set of procedures that will be used for the construction of the final vehicle.
- The team should pick the instruction that they wish to use and have justification for this.

Check point 2 - No group may progress beyond this point without this step being signed off by the instructor.

Team members can now begin organizing materials and tools to begin constructing their vehicle next class.

The fuel-cell powered Chevrolet Sequel.

VEHICLE CONSTRUCTION & PRELIMINARY TESTING

Resources & Materials Needed:
- 0.5 watt PEM fuel cells available at fuelcellstore.com
- Gear systems, assorted wheels, assorted pieces of wood, cardboard, etc for vehicle construction.

Have teams begin constructing and testing their vehicles:
- Teams begin the construction phase, using the materials provided. Teams can bring their own materials to use if approved by the instructor.
- Teams complete their design brief as appropriate.
- Teams test their vehicle and make changes as necessary.

FINAL TESTING: RACE DAY

Have teams race their fuel cell vehicles against one another to determine the fastest vehicle:
- Teams need to prepare their vehicles and have them charged and ready to race before each heat of the race.
- Teams that are not prepared when it their time to race are disqualified.

If there is time after the race, students should be completing unfinished sections of the design brief.

VEHICLE PROJECT WRAP UP

Have each team prepare a 3-5 minute presentation on their vehicle and the results of their testing for day twenty-eight.

Students can spend any extra time completing unfinished sections of the design briefs, which will be due next class period.

PROJECT PRESENTATIONS

- Teams present their vehicle and results.
- Students turn in the completed design briefs.

UNIT 8:
Biomass Energy And Bio-fuels

UNIT OBJECTIVES:
- Define a renewable fuel.
- Understand how the substitution of biodiesel fuel for petroleum diesel benefits the environment.
- Produce biodiesel fuel from waste vegetable oil.
- Understand how this fuel-making process can be adjusted to utilize waste oils from different sources.
- Perform the chemical analyses necessary to determine oil quality.
- Assess the finished products from the biodiesel reaction.
- Students should be able to discuss how issues of waste stream management can be addressed in an environmentally responsible way.

VOCABULARY

Energy	Biodiesel	Titration
Transesterification	Triglycerides	Esters
Fuel	PH	Glycerol

COLORADO MODEL CONTENT STANDARDS: BENCHMARKS FOR GRADES 9-12

1.1 Ask questions and state hypotheses, using prior scientific knowledge to help design and guide development and implementation of a scientific investigation.

1.2 Select and use appropriate technologies to gather, process, and analyze data to report information related to an investigation.

1.3 Identify major sources of error or uncertainty within an investigation. (for example: particular measuring devices and experimental procedures.

1.4 Recognizing and analyzing alternative explanations and models.

1.5 Construct and revise scientific explanations and models, using evidence, logic, and experiments that include identifying and controlling variables.

1.6 Communicate and evaluate scientific thinking that leads to particular conclusions.

2.4. Word and chemical equations are used to relate observed changes in matter to its composition and structure (for example: conservation of matter).

2.5. Quantitative relationships involved with thermal energy can be identified, measured, calculated and analyzed (for example: heat transfer in a system involving mass, specific heat, and change in temperature of matter).

2.6. Energy can be transferred through a variety of mechanisms and in any change some energy is lost as heat (for example: conduction, convection, radiation, motion, electricity, chemical bonding changes).

3.2. There is a relationship between the processes of photosynthesis and cellular respiration (for example: in terms of energy and products).

3.7. There is a cycling of matter (for example: carbon, nitrogen) and the movement and change of energy through the ecosystem (for example: some energy dissipates as heat as it is transferred through a food web).

4.4. There are costs, benefits, and consequences of natural resource exploration, development, and consumption (*for example: geosphere, biosphere, hydrosphere, atmosphere and greenhouse gas*).

4.5.There are consequences for the use of renewable and nonrenewable resources.

5.1 Print and visual media can be evaluated for scientific evidence, bias and opinion.

5.2 Identify reasons why consensus and peer review are essential to the Scientific Process.

5.3 Graphs, equations, or other models are used to analyze systems involving change and constancy (for example, comparing the geologic time scale to shorter time frames, exponential growth, a mathematical expression for gas behavior; constructing a closed ecosystem such as an aquarium).

5.4 There are cause - effect relationships within systems (for example: the effect of temperature on gas volume, effect of CO_2 level on the greenhouse effect, effects of changing in nutrients at the base of a food pyramid).

5.6 Interrelationships among science, technology and human activity lead to further discoveries that impact the world in positive and negative ways.

UNIT 8 TEACHER SCHEDULE

UNIT TIME PERIOD:
This unit is designed to take 18 45-minute class periods.

8.1. INTRODUCTION TO BIOMASS ENERGY
Day 1
Materials: NEED Biomass Transparencies, NEED Biomass Handout, What's In Your Tank? PowerPoint

8.2. BIODIESEL
Day 2: Introduction to making biodiesel
Materials: Making Biodiesel guide
Day 3: Introduction to the *Making Biodiesel Design Brief*
Materials: Making Biodiesel Design Brief
Days 4-5: The steps of making biodiesel
Days 6-13: Biodiesel synthesis and testing
Materials: See the activity guide for list of equipment, materials, & chemicals needed.
Day 14: Making soap from glycerin
Materials: Leftover glycerin from making biodiesel, set-up to heat glycerin (heat source, container, chemical hood), sodium lye, plastic soap molds.
Days 15-16: Project wrap up
Days 17-18 Project Presentations

Supplemental Activities:

8.3. *NO FOSSILS IN THIS FUEL,* MAKING ETHANOL PROJECT
Materials: See the individual activity guide for materials lists.

8.4. BUILD YOUR OWN BIOGAS GENERATOR
Materials: See the individual activity guide for materials lists.

8.1. INTRODUCTION TO BIOMASS ENERGY

Resources & Materials Needed:
- NEED Biomass Transparencies
- NEED Biomass Handout
- *What's In Your Tank?* PowerPoint

Introduce the topic of biomass energy to the students, cover the types of biomass available, and the various energy products that can be made from biomass:
- Present the NEED Biomass Transparencies, or provide them as a handout.
- Give each student the NEED Biomass Handout to read.
- Students should understand both the different energy products made from biomass (heat, electricity, vehicle fuels, ad biogas) and the types of biomass used to produce energy:
  - Wood biomass
  - Agricultural biomass
  - Solid waste
  - Landfill gas
  - Digester gas (wastewater or agricultural/livestock waste)
- Have each student describe on paper the Carbon Cycle and how biomass energy sources fit within it.
- Present the *What's In Your Tank?* PowerPoint.

8.2. BIODIESEL

INTRODUCTION TO MAKING BIODIESEL

Resources & Materials Needed:
- *Making Biodiesel* guide

This exercise introduces students to the concept of alternative fuels, and gives them an opportunity to produce their own biodiesel fuel using an analytical approach. The text of the exercise gives students a brief background in the environmental benefits of using biodiesel as a diesel substitute. The lab portion of this exercise demonstrates the basic chemistry involved in making biodiesel from vegetable oils and waste oils.

Many students have heard of biodiesel without realizing that to produce the fuel from waste vegetable oil is a fairly simple process. Seeing the process firsthand, and better yet, going through the steps from oil to fuel, enables the student to grasp the fuel making process. Included in this exercise is some basic oil analysis that is necessary to differentiate between various oils that a biodiesel producer may encounter. This is an easy exercise to set up: it requires primarily basic equipment commonly found in a high school chemistry laboratory. Interest sparked by this exercise may inspire students to become more familiar with the various aspects of renewable energy technologies.

Safety practices for handling the materials involved in producing biodiesel fuel cannot be overemphasized, especially if students attempt to synthesize biodiesel outside of class.

MAKING BIODIESEL:
Using the Making Biodiesel guide, part one:
- Review the history and background of biodiesel
- Review materials and safety: **emphasize the importance of safety when working with KOH or NaOH and methanol**
- Discuss with students the process of making biodiesel fuel

INTRODUCTION TO
THE *MAKING BIODIESEL DESIGN BRIEF*

Resources & Materials Needed:
- *Making Biodiesel Design Brief*

HAND OUT THE DESIGN BRIEF:
- Break students into groups of two to four.
- Have students put their name on the front cover and all the names of the members of their team on the instructor test data sheet.
- Assign due dates and have the students fill them into the appropriate spaces.
- Carefully go through the requirements, specifications and restrictions. Students do not write on the grading rubric on the last page. Make sure that everyone understands that deviating from or misinterpreting these requirements will affect their final grade.

REVIEW THE DESIGN BRIEF PROCEDURES:
- Go over the activity procedures and the sections of the Design Brief.
- As a homework assignment, have students read over the procedures in the design brief for making biodiesel from both fresh and waste vegetable oil.

THE STEPS OF MAKING BIODIESEL

- Introduce sections two - six of the Making Biodiesel Guide.
- Cover very carefully the steps to making fuel from both fresh and waste oil, and the titration of waste vegetable oil.

BIODIESEL SYNTHESIS AND TESTING

Resources & Materials Needed:
- *See the activity guide for list of equipment, materials, and chemicals needed.*

Each student will create biodiesel from both fresh and waste vegetable oil. Students should accomplish the following:
- Create two batches of biodiesel (fresh & waste)
- Titrate the waste vegetable to calculate the amount of NaOH needed to neutralize the waste oil.
- Complete a yield determination and wash test (Part Six)
- Critically examine and compare the two batches of oil for differences in color, clarity, and viscosity.

See the design brief for full instructions.

MAKING SOAP FROM GLYCERIN

> **Resources & Materials Needed:**
> * Leftover glycerin from making biodiesel
> * Set-up to heat glycerin (heat source, container, chemical hood)
> * Sodium lye.
> * Plastic soap molds.

Have students make soap with the glycerin byproduct of their biodiesel production:

To make soap from glycerin, heat it to 80 °C for several hours to boil off the methanol. This process must be done under a chemical hood and away from open flame. When the methanol has been removed, the liquid glycerin will stop bubbling, and the total volume of the fluid will be reduced by about 20% or more. To ensure that all of the methanol has been removed, heat to 100 °C.

For every liter of warm glycerin, add 200 mL of distilled water combined with 30 grams of sodium lye. Add the lye water to the glycerin, stir well, and pour into a plastic mold to cool. The resulting soap should cure for several weeks before use. It is effective at cutting grease on hands.

PROJECT WRAP UP

Have each group develop a 5-minute PowerPoint presentation on creating biodiesel. Presentations should include:
* Process of using fresh oil
* Process of using waste oil:
    o Where did the oil come from?
    o Titration results and amount of NaOH needed
* Process of washing the fuel
* Yield & wash test results

You may also want to give the students time to work individually on completing their design briefs.

PROJECT PRESENTATIONS

* Teams present their PowerPoints to the class.
* Students turn in the design briefs.

Supplemental Activities

You may want to use these activities as additions or alternatives to the *Making Biodiesel Design Brief*:

8.3. NO FOSSILS IN THIS FUEL, MAKING ETHANOL PROJECT:

Resources & Materials Needed:
- See the project activity guide for materials lists.

By the GM Environmental Science Club. This activity has students use yeast and corn syrup to create ethanol.

See the activity guide for instructions and materials.

Front Range Energy's ethanol plant in Windsor, CO.

8.4. BUILD YOUR OWN BIOGAS GENERATOR:

Resources & Materials Needed:
- See the project activity guide for materials lists.

By the Pembina Institute.
This activity has students construct a biogas-producing manure digester from an 18L plastic water container. This is a particularly good activity for schools in areas with farms and ranches.

See the activity guide for instructions and materials.

UNIT 9:
Geothermal Energy

UNIT OBJECTIVES:
- Increase the students awareness of different types of geothermal energy capture.
- Apply knowledge of energy use and efficiency to real life situations.
- Students should be able to discuss what factors enhance or hinder our attempts at gathering geothermal energy.

VOCABULARY:

Geothermal energy	Geyser	Flashed steam plants
Hot dry rock	Geothermal heat pump	Dry steam plants
Steam turbine	Ring of Fire	Binary power plants

COLORADO MODEL CONTENT STANDARDS: BENCHMARKS FOR GRADES 9-12

1.1 Ask questions and state hypotheses, using prior scientific knowledge to help design and guide development and implementation of a scientific investigation.

1.2 Select and use appropriate technologies to gather, process, and analyze data to report information related to an investigation.

1.3 Identify major sources of error or uncertainty within an investigation. (for example: particular measuring devices and experimental procedures).

1.4 Recognizing and analyzing alternative explanations and models.

1.5 Construct and revise scientific explanations and models, using evidence, logic, and experiments that include identifying and controlling variables.

1.6 Communicate and evaluate scientific thinking that leads to particular conclusions.

2.5. Quantitative relationships involved with thermal energy can be identified, measured, calculated and analyzed (for example: heat transfer in a system involving mass, specific heat, and change in temperature of matter).

2.6. Energy can be transferred through a variety of mechanisms and in any change some energy is lost as heat (for example: conduction, convection, radiation, motion, electricity, chemical bonding changes).

4.1.The earth's interior has a composition and structure.

4.2.The theory of plate tectonics helps to explain relationships among earthquakes, volcanoes, midocean ridges, and deep-sea trenches.

4.4. There are costs, benefits, and consequences of natural resource exploration, development, and consumption (*for example: geosphere, biosphere, hydrosphere, atmosphere and greenhouse gas*).

4.5.There are consequences for the use of renewable and nonrenewable resources.

5.3 Graphs, equations, or other models are used to analyze systems involving change and constancy (for example, comparing the geologic time scale to shorter time frames, exponential growth, a mathematical expression for gas behavior; constructing a closed ecosystem such as an aquarium).

5.4 There are cause - effect relationships within systems (for example: the effect of temperature on gas volume, effect of CO_2 level on the greenhouse effect, effects of changing in nutrients at the base of a food pyramid).

5.6 Interrelationships among science, technology and human activity lead to further discoveries that impact the world in positive and negative ways.

Geothermal hot springs pool in Glenwood Springs, CO.

UNIT TIME PERIOD:
This unit is designed to take 10 45-minute class periods.

9.1. INTRODUCTION TO GEOTHERMAL ENERGY
Day 1
Materials: Geothermal Heat Pump PowerPoint, copies of the NEED, Energy for Keeps, and DOE geothermal handouts.

9.2. STEAM TURBINE PROJECT
Days 2-4
Materials:
- *aluminum pie pans*
- *Metal funnel, 4 inches (10 cm) in diameter*
- *Scissors, compasses (for drawing circles), rulers, pencils, push pins,*
- *Several plastic straws (the long soda type is best)*
- *Small thin washers (optional)*
- *Small cooking pots, no bigger than 5 or 6 inches (13-15 cm) in diameter*
- *Hot plate(s) or other heat source(s)*
- *Oven mitts, towels for clean-up*
- *Source(s) of falling water, such as a faucet and sink, or a large jug or bottle of water and a bucket or tub*

9.3. GEOTHERMAL BUILDING SYSTEMS
Day 5
Materials: Geothermal Home Heating PowerPoint, a copy of What is Geothermal Energy? for each student.

9.4. GEOTHERMAL WORKBOOK EXERCISES
Days 6-8
Materials: See the Geothermal Energy Workbook activity guides for instructions and needed materials.

9.5. MODEL GEYSER PROJECT
Days 9-10
Materials: See the Model Geyser project activity guide for geyser-construction instructions.

9.1. INTRODUCTION TO GEOTHERMAL ENERGY

Resources & Materials Needed:
- *Geothermal Heat Pump* PowerPoint
- Copies of the NEED, *Energy for Keeps*, and DOE geothermal handouts.

Present the *Geothermal Heat Pump* PowerPoint, by Warren Thomas of Oak Ridge National Laboratory. The presentation covers:
- An overview of the technology
- Different types geothermal of systems
- Important factors in design
- System economics

After the presentation, give students copies and assign for reading the DOE *Geothermal Fact Sheet*, *Energy for Keeps-Geothermal*, and NEED geothermal handouts.

9.2. STEAM TURBINE PROJECT

Resources & Materials Needed:
- Aluminum pie pans
- Metal funnel, 4 inches (10 cm) in diameter
- Scissors, compasses (for drawing circles), rulers, pencils, push pins,
- Several plastic straws (the long soda type is best)
- Small thin washers (optional)
- Small cooking pots, no bigger than 5 or 6 inches (13-15 cm) in diameter
- Hot plate(s) or other heat source(s)
- Oven mitts, towels for clean-up
- Source(s) of falling water, such as a faucet and sink, or a large jug or bottle of water and a bucket or tub

Before beginning the Steam Turbine project:
- Present the *Geothermal Energy* PowerPoint, by Chuck Kutscher, NREL. The presentation provides additional information on the use and availability of geothermal information, and on current and future geothermal technologies.
- Hand out *How a Steam Generator Works*, from *Energy for Keeps.*

MAKING A MODEL STEAM TURBINE:
This project, from *Energy for Keeps*, has students experiment with the mechanics of using steam to generate electricity. **See the activity guide PDF for details, materials, and instructions.**

9.3. GEOTHERMAL BUILDING SYSTEMS

Resources & Materials Needed:
- *Geothermal Home Heating* PowerPoint
- A copy of *What is Geothermal Energy?* for each student.

- If possible, try to arrange for a guest speaker from a local geothermal company to visit and talk about their work and the kind of systems they install.
- If you're not able to arrange for a guest speaker, present the *Geothermal Home Heating* PowerPoint.
- Hand out and assign as reading *What is Geothermal Energy?* by Mary H. Dickson and Mario Fanelli.
- Lead a review and student discussion of *What is Geothermal Energy?*

9.4. GEOTHERMAL WORKBOOK EXERCISES

Resources & Materials Needed:
* See the *Geothermal Energy Workbook* activity guides for instructions and needed materials.

Complete the activities in sections 2 and 4 of the *Geothermal Energy Workbook,* by the Geothermal Energy Office. **See the activity guides for instructions and needed materials.**

9.5. MODEL GEYSER PROJECT

Resources & Materials Needed:
* See the *Model Geyser* project activity guide for geyser-construction instructions.

The *Model Geyser* project, by Clint Sprott, has students construct a small, functioning geyser with a water-filled tube (such as one made of pyrex glass), a Bunsen burner and a catch basin to collect and recycle the water.
See the activity guide for full instructions.

UNIT 10:
Hydropower

UNIT OBJECTIVES:
- Increase the students awareness of different types of water energy capture.
- Apply knowledge of energy use and efficiency to real life situations.
- Students should be able to discuss what the methods used to collect energy from moving water.

VOCABULARY:

Head	OTEC	Barrage
Flow	Tidal energy	Sluice
Hydropower	Wave energy	Reservoir

COLORADO MODEL CONTENT STANDARDS: BENCHMARKS FOR GRADES 9-12

1.1 Ask questions and state hypotheses, using prior scientific knowledge to help design and guide development and implementation of a scientific investigation.

1.2 Select and use appropriate technologies to gather, process, and analyze data to report information related to an investigation.

1.3 Identify major sources of error or uncertainty within an investigation (for example: particular measuring devices and experimental procedures).

1.4 Recognizing and analyzing alternative explanations and models.

1.5 Construct and revise scientific explanations and models, using evidence, logic, and experiments that include identifying and controlling variables.

1.6 Communicate and evaluate scientific thinking that leads to particular conclusions.

2.4. Word and chemical equations are used to relate observed changes in matter to its composition and structure (for example: conservation of matter).

2.5. Quantitative relationships involved with thermal energy can be identified, measured, calculated and analyzed (for example: heat transfer in a system involving mass, specific heat, and change in temperature of matter).

2.6. Energy can be transferred through a variety of mechanisms and in any change some energy is lost as heat (for example: conduction, convection, radiation, motion, electricity, chemical bonding changes).

4.4. There are costs, benefits, and consequences of natural resource exploration, development, and consumption (*for example: geosphere, biosphere, hydrosphere, atmosphere and greenhouse gas*).

4.5.There are consequences for the use of renewable and nonrenewable resources.

5.1 Print and visual media can be evaluated for scientific evidence, bias and opinion.

5.2 Identify reasons why consensus and peer review are essential to the Scientific Process.

5.3 Graphs, equations, or other models are used to analyze systems involving change and constancy (for example, comparing the geologic time scale to shorter time frames, exponential growth, a mathematical expression for gas behavior; constructing a closed ecosystem such as an aquarium).

5.4 There are cause - effect relationships within systems (for example: the effect of temperature on gas volume, effect of CO_2 level on the greenhouse effect, effects of changing in nutrients at the base of a food pyramid).

5.6 Interrelationships among science, technology and human activity lead to further discoveries that impact the world in positive and negative ways.

UNIT TIME PERIOD:
This unit is designed to take 6 45-minute class periods.

10.1. INTRODUCTION TO HYDROELECTRIC ENERGY
Day 1
Materials: *A copy of the NEED Hydropower handout for each student, a copy of Calculating the Potential Energy of Flowing Water for each student.*

10.2. CALCULATING THE ENERGY IN WATER—FIELD TRIP
Day 2
Materials:
- Transportation to stream, if needed.
- Meter sticks, matchsticks, 30-meter (100-ft) tape measure, stopwatch
- Topographic map of area
- Copies of student sheets from Calculating the Energy in Water activity guide.

10.3. GENERATE YOUR OWN HYDROPOWER ACTIVITY
Days 3-5
Materials:
- *Copies of Generate Your Own Hydropower activity guide.*
- *2 alligator clips (optional)*
- *Small spool magnetic wire (#28 or finer, insulated)*
- *2 cardboard or masonite rectangles (about 5" x 7"), 8 small paper cups.*
- *Compass, glue, electrical tape, 2 1-inch nails, 2 3-inch nails*
- *1-inch bar magnet*
- *2 1-1/2"x4" metal strips cut from tin can*
- *Germanium diode (for example, type 1N34A)*
- *Soldering iron (optional), solder (optional)*
- *3x5" wood block, round tinker toy, 8 3" tinker toy spokes*

10.4. OCEAN POWER
Day 6
Materials: *Energy for Keep: Ocean Power handouts.*

10.1. INTRODUCTION TO HYDROELECTRIC ENERGY

Resources & Materials Needed:
- A copy of the *NEED Hydropower* handout for each student.
- A copy of *Calculating the Potential Energy of Flowing Water* for each student.

BASICS OF HYDROPOWER
Hand out copies of the NEED *Hydropower* handout. If you have *Energy for Keeps*, you can also provide them with copies of the hydropower chapter.

Review and discuss the handout with students. You may want to talk about the different scales of hydropower systems, and have students discuss the pros (renewable energy, water supply, and flood control) and cons (negative ecological and social impacts) of large dams. Examples for discussion could include the Three Gorges Dam on the Yangtze River in China, the Glen Canyon dam in Arizona, or the history of the Yosemite and Hetch Hetchy Valleys in California.

CALCULATING THE ENERGY OF WATER
Hand out the *Calculating the Potential Energy of Flowing Water* project guide (from the *High School Energy Sourcebook*, by the Tennessee Valley Authority). Prepare students for the activity and review the activity procedures. See the activity guide for full instructions.

10.2. CALCULATING THE ENERGY IN WATER FIELD TRIP

Resources & Materials Needed:
- Transportation to stream, if needed.
- Meter sticks, matchsticks, 30-meter (100-ft) tape measure, stopwatch
- Topographic map of area
- Copies of student sheets from *Calculating the Energy in Water* activity guide.

Take students to a nearby stream or river, and have them calculate the potential energy of the stream flow. **See the activity guide for full instructions.**

You may want to take time on day three to review the activity and results.

10.3. GENERATE YOUR OWN HYDROPOWER

Resources & Materials Needed:
- Copies of *Generate Your Own Hydropower* activity guide.
- 2 alligator clips (optional)
- small spool magnetic wire (#28 or finer, insulated)
- 2 cardboard or masonite rectangles (about 5" x 7"), 8 small paper cups.
- compass, glue, electrical tape, 2 1-inch nails, 2 3-inch nails
- 1-inch bar magnet
- 2 1-1/2"x4" metal strips cut from tin can
- germanium diode (for example, type 1N34A)
- soldering iron (optional), solder (optional)
- 3x5" wood block, round tinker toy, 8 3" tinker toy spokes

The *Generate Your Own Hydropower* activity, from the *High School Energy Sourcebook*, by the Tennessee Valley Authority, has students construct a model hydropower generator:
- Divide students into pairs.
- Hand out the *Generate Your Own Hydropower* activity (from the *High School Energy Sourcebook*, by the Tennessee Valley Authority).
- Review the activity procedures with students and have them begin working on their projects.

See the activity guide for details and full instructions.

10.4. OCEAN POWER

Resources & Materials Needed:
* A copy of the *Energy for Keeps: Ocean Power* handout for each student.

Give each student a copy of the *Energy for Keeps: Ocean Power* handout, and assign as reading homework. Students should come prepared to discuss in the next class.

Ocean power is an emerging technology thought to have much potential. Review the assigned reading and lead a discussion on the potential for this technology. You may want to highlight some of the current companies who are developing ocean power technologies and projects:
* Ocean Power Technology, www.oceanpowertechnologies.com
* Finavera Renewables, www.finavera.com
* Oceanlinx, www.oceanlinx.com
* Pelamis Wave Power, www.pelamiswave.com
* ORECon, www.orecon.com

An Ocean Power Technology PowerBouy.

UNIT 11:
Final
Project

UNIT OBJECTIVES:
- Students should be able to demonstrate their understanding of a specific renewable technology.
- To professionally present their research on a specific technology to the class.

COLORADO MODEL CONTENT STANDARDS: BENCHMARKS FOR GRADES 9-12
1.6 Communicate and evaluate scientific thinking that leads to particular conclusions.

4.4. There are costs, benefits, and consequences of natural resource exploration, development, and consumption (*for example: geosphere, biosphere, hydrosphere, atmosphere and greenhouse gas*).

4.5. There are consequences for the use of renewable and nonrenewable resources.

5.1 Print and visual media can be evaluated for scientific evidence, bias and opinion.

5.2 Identify reasons why consensus and peer review are essential to the Scientific Process.

5.3 Graphs, equations, or other models are used to analyze systems involving change and constancy (for example, comparing the geologic time scale to shorter time frames, exponential growth, a mathematical expression for gas behavior; constructing a closed ecosystem such as an aquarium).

5.4 There are cause - effect relationships within systems (for example: the effect of temperature on gas volume, effect of CO_2 level on the greenhouse effect, effects of changing in nutrients at the base of a food pyramid).

5.5 Scientific knowledge changes and accumulates over time; usually the changes that take place are small modifications of prior knowledge about major shifts in the scientific view of how the world works do occur.

5.6 Interrelationships among science, technology and human activity lead to further discoveries that impact the world in positive and negative ways.

UNIT 11 TEACHER SCHEDULE

UNIT TIME PERIOD:
This unit is designed to take 10 45-minute class periods.

There are two options for the final project:

11.1. FINAL OPTION 1: STUDENT-CREATED DESIGN BRIEFS
Days 1-10
Materials: *Final Project handouts, computers for research and writing.*
Finals Period: Final presentations

11.2. FINAL OPTION 2: RESEARCH PAPER
Days 1-10
Materials: *Final Project handouts, computers for research and writing.*
Finals Period: Final presentations

11.1. FINAL OPTION 1: STUDENT-CREATED DESIGN BRIEFS

Resources & Materials Needed:
- Computers for research and writing.
- Any materials necessary for the student-created design briefs (students may need to provide their own materials).
- *Final Project Option 1* handout.

- Students will develop and test their own Design Briefs based on a renewable energy source.
- Students will then present the projects to the class and, if possible members of school staff and the community.

See the final project requirement handout text below for additional information.

Final Project Requirements

1. Create your own Design Brief
   - Choose a specific portion a renewable energy source
   - Use a design brief we have done in class as a template.
   - Examples:
     - Use the BTU or Bust or Hydrogen Design Brief for projects
     - Use the Biodiesel Design Brief and Instructions as a template for experiments
     - You must test your experiment and brief and record the results for the presentation (Must record data for at least 4 criteria)

   - The following areas need to be addressed;
     - Cover page
     - Requirements & Idea Development
     - Research Questions (at least 5 no more than 10)
     - Visual ideas and Final Design
     - Construction or experiment steps
     - 7 resources of technology
     - Preliminary Testing & Results
     - Final Testing
     - Group Member Assessment
     - Grading Rubric
     - Bibliography
     - Assessment of project group members

2. Design an experiment or activity related to the topic.
   - Project or experiment must be topic specific.
   - What is the purpose or goal of the project or experiment?
     - For projects: prepare to have an example of the activity
     - For experiments: plan to do the experiment during your presentation

3. PowerPoint Presentation
   - Administration, staff, and NREL employees will be attending
   - 15-20 minutes (50 pt deduction for every minute outside the range)
   - All members must participate in the presentation
   - Professional Dress is required
   - The following is required;
     - Topic introduction, description, and background information
     - Present entire design brief
     - Present activity or experiment
     - Present results of the testing of your brief and experiment
     - Acknowledgments
     - Questions

FINAL PRESENTATIONS

Student groups give their PowerPoint presentations.

Students submit their research papers.

11.2. FINAL OPTION 2: RESEARCH PAPER

Resources & Materials Needed:
- Computers for research and writing.
- *Final Project Option 2* handout.

- Students will each research and write a 3-5 page paper on a topic area from the course.
- Once students have completed their papers, divide them into group of no more than four, organized by topic area.
- Have groups select as a presentation topic a specific technology from their topic area for example:
  - If the group topic is solar, they might present on concentrated solar collector systems.
  - If their topic is biomass, they might present on landfill-gas to energy technology.

See the final project requirement handout text below for additional information.

Final Research Paper Requirements

You must complete your paper and work with your team to present your renewable energy source during finals.

1. *Select 1 topic from the following list:*

   Solar   Biomass
   Wind and Hydroelectric   Hydrogen Fuel Cells
   Green Building and Sustainability   Renewable Transportation
   Geothermal   Efficiency & Conservation

2. *Your goal is to research the topic you choose and write a comprehensive paper. Your paper must be 3-5 pages and must include but is not limited to the following subject areas.*

   Introduction   Costs and Availability
   History   Current Research
   Role in the Energy Picture   Future Research
   Important People   Implementation and Integration
   Operation (How it Works)   Conclusion

   Your paper must be properly referenced (reference page), not included in the 3-5 page requirement

---

The amount of work that you put into this and the quality of the research done will reflect on your final grade for this paper.

3. *After you have written your paper, you will work in groups of no more than four to put together a PowerPoint Presentation. As a group, you must pick one aspect of your energy source and give a detailed description of that aspect. For example, if your topic was solar, you might do your presentation on Solar Hot Water Systems. Your presentation must include but is not limited to the following:*

Introduction

History

Role in the Energy Picture

Important People

Operation (How it Works)

Costs and Availability

Current Research

Future Research

Implementation and Integration

Conclusion

Note that the presentation requirements mimic your paper requirements. As the group comes together to put the presentation together AFTER the individual papers are written you should have a lot of good information but more research will be required. Remember that the paper is on the general topic and the presentation is on a specific aspect of the topic. Everyone in the group MUST come to a consensus as to the topic and then participate in the presentation. You are the experts on your topic. Professional dress is expected.

FINAL PRESENTATIONS

Student groups give their PowerPoint presentations.

Students submit their research papers.

Appendix

NATIONAL SCIENCE EDUCATION STANDARDS ALIGNMENT

National Science Education Standard	Unit										
	1	2	3	4	5	6	7	8	9	10	11
Content Standard A: Science as Inquiry											
Abilities Necessary to Do Scientific Inquiry											
• Identify questions and concepts that guide scientific investigations	✓	✓	✓	✓	✓	✓	✓	✓	✓	✓	✓
• Design and conduct scientific investigations.	✓	✓	✓	✓	✓	✓	✓	✓	✓	✓	
• Use technology and mathematics to improve investigations and communications.	✓	✓	✓	✓	✓	✓	✓	✓	✓	✓	✓
• Formulate and revise scientific explanations and models using logic and evidence.	✓	✓	✓	✓	✓	✓	✓	✓	✓	✓	✓
• Recognize and analyze alternative explanations and models.	✓	✓	✓	✓	✓	✓	✓	✓	✓	✓	✓
• Communicate and defend a scientific argument.	✓	✓	✓	✓	✓	✓	✓	✓	✓	✓	✓
Understandings about Scientific Inquiry											
• Scientists usually inquire about how physical, living, or designed systems function. Conceptual principles and knowledge guide scientific inquiries. Historical and current scientific knowledge influence the design and interpretation of investigations and the evaluations of proposed explanations made by other scientists.	✓										
• Scientists conduct investigations for a wide variety of reasons. For example, they may wish to discover new aspects of the natural world, explain recently observed phenomena, or test the conclusions of prior investigations or the predictions of current theories.	✓										
• Scientists rely on technology to enhance the gathering and manipulation of data. New techniques and tools provide new evidence to guide inquiry and new methods to gather data, thereby contributing to the advance of science. The accuracy and precision of the data, and therefore the quality of the exploration, depends on the technology used.	✓	✓	✓	✓	✓	✓	✓	✓	✓	✓	✓
• Mathematics is essential in scientific inquiry. Mathematical tools and models guide and improve the posing of questions, gathering data, constructing explanations and communicating results.	✓	✓	✓	✓	✓	✓	✓	✓	✓	✓	✓

National Science Education Standard	1	2	3	4	5	6	7	8	9	10	11
• Scientific explanations must adhere to criteria such as: a proposed explanation must be logically consistent; it must abide by the rules of evidence; it must be open to questions and possible modification; and it must be based on historical and current scientific knowledge.	✓	✓	✓	✓	✓	✓	✓	✓	✓	✓	✓
• Results of scientific inquiry—new knowledge and methods—emerge from different types of investigations and public communication among scientists. In communicating and defending the results of scientific inquiry, arguments must be logical and demonstrate connections between natural phenomena, investigations, and the historical body of scientific knowledge. In addition, the methods and procedures that scientists used to obtain evidence must be clearly reported to enhance opportunities for further investigation.	✓	✓	✓	✓	✓	✓	✓	✓	✓	✓	✓
Content Standard B: Physical Science											
Structure of atoms											
• Radioactive isotopes are unstable and undergo spontaneous nuclear reactions, emitting particles and/or wavelike radiation. The decay of any one nucleus cannot be predicted, but a large group of identical nuclei decay at a predictable rate. This predictability can be used to estimate the age of materials that contain radioactive isotopes.											
Chemical reactions											
• Chemical reactions may release or consume energy. Some reactions such as the burning of fossil fuels release large amounts of energy by losing heat and by emitting light. Light can initiate many chemical reactions such as photosynthesis and the evolution of urban smog.	✓					✓	✓	✓			
Conservation of energy and increase in disorder											
• The total energy of the universe is constant. Energy can be transferred by collisions in chemical and nuclear reactions, by light waves and other radiations, and in many other ways. However, it can never be destroyed. As these transfers occur, the matter involved becomes steadily less ordered.	✓	✓	✓	✓	✓	✓	✓	✓	✓	✓	✓

National Science Education Standard

National Science Education Standard	1	2	3	4	5	6	7	8	9	10	11
• All energy can be considered to be either kinetic energy, which is the energy of motion; potential energy, which depends on relative position; or energy contained by a field, such as electromagnetic waves.	✓										
• Heat consists of random motion and the vibrations of atoms, molecules, and ions. The higher the temperature, the greater the atomic or molecular motion.	✓										
• Everything tends to become less organized and less orderly over time. Thus, in all energy transfers, the overall effect is that the energy is spread out uniformly. Examples are the transfer of energy from hotter to cooler objects by conduction, radiation, or convection and the warming of our surroundings when we burn fuels.	✓		✓	✓	✓	✓	✓		✓	✓	✓
Interactions of energy and matter											
• In some materials, such as metals, electrons flow easily, whereas in insulating materials such as glass they can hardly flow at all. Semi-conducting materials have intermediate behavior. At low temperatures some materials become superconductors and offer no resistance to the flow of elections.	✓		✓	✓							
Content Standard C: Life Science											
Interdependence of organisms											
• The atoms and molecules on the earth cycle among the living and nonliving components of the biosphere.	✓							✓			
• Energy flows through ecosystems in one direction, from photosynthetic organisms to herbivores to carnivores and decomposers.	✓							✓			
• Organisms both cooperate and compete in ecosystems. The interrelationships and interdependencies of these organisms may generate ecosystems that are stable for hundreds or thousands of years.	✓										
• Living organisms have the capacity to produce populations of infinite size, but environments and resources are finite. This fundamental tension has profound effects on the interactions between organisms.	✓										

National Science Education Standard

National Science Education Standard	1	2	3	4	5	6	7	8	9	10	11
• Human beings live within the world's ecosystems as a result of population growth, technology, and consumption. Human destruction of habitats through direct harvesting, pollution, atmospheric changes, and other factors is threatening current global stability, and if not addressed, ecosystems will be irreversibly affected.	✓										
Content Standard D: Earth and Space Science											
Energy in the Earth System											
• Earth systems have internal and external sources of energy, both of which create heat. The sun is the major external source of energy. Two primary sources of internal energy are the decay of radioactive isotopes and the gravitational energy from the earth's original formation.	✓			✓					✓		
• The outward transfer of earth's internal heat drives convection circulation in the mantle that propels the plates comprising earth's surface across the face of the globe.									✓		
• Heating of earth's surface and atmosphere by the sun drives convection within the atmosphere and oceans, producing winds and ocean currents.					✓						
• Global climate is determined by energy transfer from the sun at or near the earth's surface. This energy transfer is influenced by dynamic processes such as cloud cover and the earth's rotation, and static conditions such as the position of mountain ranges and oceans.	✓			✓	✓						
Geochemical Cycles											
• The earth is a system containing essentially a fixed amount of each stable chemical atom or element. Each element can exist in several different chemical reservoirs. Each element on earth moves among reservoirs in the solid earth, oceans, atmosphere, and organisms as part of geochemical cycles.	✓							✓			
• Movement of matter between reservoirs is driven by the earth's internal and external sources of energy. These movements are often accompanied by a change in the physical and chemical properties of the matter. Carbon, for example, occurs in carbonate rocks such as limestone, in the atmosphere as carbon dioxide gas, in water as dissolved carbon dioxide, and in all organisms as complex molecules that control the chemistry of life.	✓							✓			

National Science Education Standard

National Science Education Standard	1	2	3	4	5	6	7	8	9	10	11
The Origin and Evolution of the Universe											
• Stars produce energy from nuclear reactions, primarily the fusion of hydrogen to form helium. These and other processes in stars have led to the formation of all the other elements.											
Content Standard E: Science and Technology											
Abilities of Technological Design											
• Identify a Problem or Design and Opportunity			✓	✓	✓	✓	✓	✓	✓	✓	✓
• Propose Designs and Choose Between Alternative Solutions			✓	✓	✓	✓	✓	✓	✓	✓	
• Implement a Proposed Solution			✓	✓	✓	✓	✓	✓	✓	✓	
• Evaluate the Solution and Its Consequences			✓	✓	✓	✓	✓	✓	✓	✓	
• Communicate the Problem, Process, and Solution.			✓	✓	✓	✓	✓	✓	✓	✓	✓
Understandings about Science and Technology											
• Scientists in different disciplines ask different questions, use different methods of investigation, and accept different types of evidence to support their explanations. Many scientific investigations require the contributions of individuals from different disciplines, including engineering. New disciplines of science, such as geophysics and biochemistry often emerge at the interface of two older disciplines.											
• Science often advances with the introduction of new technologies. Solving technological problems often results in new scientific knowledge. New technologies often extend the current levels of scientific understanding and introduce new areas of research.	✓	✓	✓	✓	✓	✓	✓	✓	✓	✓	✓
• Creativity, imagination and a good knowledge base are all required in the work of science and engineering.	✓	✓	✓	✓	✓	✓	✓	✓	✓	✓	✓

National Science Education Standard

	1	2	3	4	5	6	7	8	9	10	11
• Science and technology are pursued for different purposes. Scientific inquiry is driven by the desire to understand the natural world, and technological design is driven by the need to meet human needs and solve human problems. Technology, by its nature, has a more direct effect on society than science because its purpose is to solve human problems, help humans adapt, and fulfill human aspirations. Technological solutions may create new problems. Science by its nature, answers questions that may or may not directly influence humans. Sometimes scientific advances challenge people's beliefs and practical explanations concerning various aspects of the world.	✓	✓	✓	✓	✓	✓	✓	✓	✓	✓	✓
• Technological knowledge is often not made public because of patents and the financial potential of the idea or invention. Scientific knowledge is made public through presentations at professional meetings and publications in scientific journals.											

Content Standard F: Science in Personal and Social Perspectives

Personal and Community Health

	1	2	3	4	5	6	7	8	9	10	11
• Hazards and the potential for accidents exist. Regardless of the environment, the possibility of injury, illness, disability, or death may be present. Humans have a variety of mechanisms—sensory, motor, emotional, social, and technological—that can reduce and modify hazards.	✓	✓	✓	✓	✓	✓	✓	✓	✓	✓	✓

Population Growth

	1	2	3	4	5	6	7	8	9	10	11
• Populations grow or decline through the combined effects of births and deaths, and through emigration and immigration. Populations can increase through linear or exponential growth, with effects on resource use and environmental pollution.	✓										
• Populations can reach limits to growth. Carrying capacity is the maximum number of individuals that can be supported in a given environment. The limitation is not the availability of space, but the number of people in relation to resources and the capacity of earth systems to support human beings. Changes in technology can cause significant changes, either positive or negative, in carrying capacity.	✓										

National Science Education Standard	1	2	3	4	5	6	7	8	9	10	11
Natural Resources											
• Human populations use resources in the environment in order to maintain and improve their existence. Natural resources have been and will continue to be used to maintain human populations.	✓										
• The earth does not have infinite resources: increasing human consumption places severe stress on the natural processes that renew some resources, and it depletes those resources than cannot be renewed.	✓										
• Humans use many natural systems as resources. Natural systems have the capacity to reuse waste, but that capacity is limited. Natural systems can change to an extent that exceeds the limits of organisms to adapt naturally or humans to adapt technologically.	✓										
Environmental Quality											
• Natural ecosystems provide an array of basic processes that affect humans. Those processes include maintenance of the quality of the atmosphere, generation of soils, control of the hydrologic cycle, disposal of wastes, and recycling of nutrients. Humans are changing many of these basic processes, and the changes may be detrimental to humans.	✓										
• Materials from human societies affect both physical and chemical cycles of the earth.	✓										
• Many factors influence environmental quality. Factors that students might investigate include population growth, resource use, population distribution, over consumption, the capacity of technology to solve problems, poverty, the role of economic, political, and religious views, and different ways humans view the earth.	✓										

Revision date: 6/1/08

National Science Education Standard

National Science Education Standard	1	2	3	4	5	6	7	8	9	10	11
Natural and Human-induced Hazards											
• Normal adjustments of earth may be hazardous for humans. Humans live at the interface between the atmosphere driven by solar energy and the upper mantle where convection creates changes in the earth's solid crust. As societies have grown, become stable, and come to value aspects of the environment, vulnerability to natural processes of change has increased.	✓										
• Human activities can enhance potential for hazards. Acquisition of resources, urban growth, and waste disposal can accelerate rates of natural change.	✓										
• Some hazards, such as earthquakes, volcanic eruptions, and severe weather, are rapid and spectacular. But there are slow and progressive changes that also result in problems for individuals and societies. For example, change in stream channel position, erosion of bridge foundations, sedimentation in lakes and harbors, coastal erosions, and continuing erosion and wasting of soil and landscapes can all negatively affect society.	✓										
• Natural and human-induced hazards present the need for humans to assess potential danger and risk. Many changes in the environment designed by humans bring benefits to society, as well as cause risks. Students should understand the costs and trade-offs of various hazards—ranging from those with minor risk to a few people to major catastrophes with major risk to many people. The scale of events and the accuracy with which scientists and engineers can (and cannot) predict events are important considerations.	✓										
Content Standard G: History and Nature of Science											
Science as a Human Endeavor											
• Individuals and teams have contributed and will continue to contribute to the scientific enterprise. Doing science or engineering can be as simple as an individual conducting field studies or as complex as hundreds of people working on a major scientific question or technological problem. Pursuing science as a career or as a hobby can be both fascinating and intellectually rewarding.		✓	✓	✓	✓	✓	✓	✓	✓	✓	✓

National Science Education Standard	1	2	3	4	5	6	7	8	9	10	11
• Scientists have ethical traditions. Scientists value peer review, truthful reporting about the methods and outcomes of investigation, and making public the results of work. Violations of such norms do occur, but scientists responsible for such violations are censured by their peers.	✓	✓	✓	✓	✓	✓	✓	✓	✓	✓	✓
• Scientists are influenced by societal, cultural, and personal beliefs and ways of viewing the world. Science is not separate from society but rather science is a part of society.	✓	✓	✓	✓	✓	✓	✓	✓	✓	✓	✓
Nature of Scientific Knowledge											
• Science distinguishes itself from other ways of knowing and from other bodies of knowledge through the use of empirical standards, logical arguments, and skepticism, as scientists strive for the bet possible explanations about the natural world.	✓	✓	✓	✓	✓	✓	✓	✓	✓	✓	✓
• Scientific explanations must meet certain criteria. First and foremost, they must be consistent with experimental and observational evidence about nature, and must make accurate predictions, when appropriate, about systems being studied. They should also be logical, respect the rules of evidence, be open to criticism, report methods and procedures, and make knowledge public. Explanations on how the natural world changes base on myths, personal beliefs, religious values, mystical inspiration, superstition or authority may be personally useful and socially relevant, but they are not scientific.	✓	✓	✓	✓	✓	✓	✓	✓	✓	✓	
• Because all scientific ideas depend on experimental and observational confirmation, all scientific knowledge is, in principle, subject to change as new evidence becomes available. The core ideas of science such as the conservation of energy or the laws of motion have been subjected to a wide variety of confirmations and are therefore unlikely to change in the areas in which they have been tested. In areas where data or understanding are incomplete, such as the details of human evolution or questions surrounding global warming, new data may well lead to changes in current ideas or resolve current conflicts. In situations where information is still fragmentary, it is normal for scientific ideas to be incomplete, but this is also where the opportunity for making advances may be greatest.	✓	✓	✓	✓	✓	✓	✓	✓	✓	✓	

National Science Education Standard	1	2	3	4	5	6	7	8	9	10	11
Historical Perspectives											
• In history, diverse cultures have contributed scientific knowledge and technologic inventions. Modern science began to evolve rapidly in Europe several hundred years ago. During the past two centuries, it has contributed significantly to the industrialization of Western and non-Western cultures. However, other, non-European cultures have developed scientific ideas and solved human problems through technology.	✓	✓	✓	✓	✓	✓	✓	✓	✓	✓	✓
• Usually, changes in science occur as small modifications in extant knowledge. The daily work of science and engineering results in incremental advances in our understanding of the world and our ability to meet human needs and aspirations. Much can be learned about the internal workings of science and the nature of science from study of individual scientists, their daily work, and their efforts to advance scientific knowledge in their area of study.	✓	✓	✓	✓	✓	✓	✓	✓	✓	✓	✓
• Occasionally, there are advances in science and technology that have important and long-lasting effects on science and society. Examples of such advances include the following: *Copernican revolution, Newtonian mechanics, Relativity, Geologic time scale, Plate tectonics, Atomic theory, Nuclear physics, Biological evolution, Germ theory, Industrial revolution, Molecular biology, Information and communication, Quantum theory, Galactic universe, Medical and health technology.*	✓	✓	✓	✓	✓	✓	✓	✓	✓	✓	✓
• The historical perspective of scientific explanations demonstrates how scientific knowledge changes by evolving over time, almost always building on earlier knowledge.	✓	✓	✓	✓	✓	✓	✓	✓	✓	✓	✓

www.ingramcontent.com/pod-product-compliance
Lightning Source LLC
Chambersburg PA
CBHW080827180526
45168CB00006B/2604